Safety Manual for Indian Laboratories

Dr Ajay Kumar Gupta
Pritanshu

DEPARTMENT OF BIOTECHNOLOGY

MAHARISHI MARKANDESHWAR UNIVERSITY

MULLANA- 133 207 (HARYANA)

ACKNOWLEDGEMENTS

All works and researches done by worthy scientists are dully cited and have formally referred in this book. Should any authors have any concern are most welcome to contact the authors. We shall try to rectify the same with utmost care and obligations.

Dr. Ajay Kumar Gupta
Pritanshu

Preface

This lab manual is designed to ensure safe working in laboratories. Content is basically according to Indian conditions. The manual puts emphasis on adherence to lab rules and regulations. There is detailed information regarding safe handling of lab equipments and chemicals. This lab manual answers all the basic questions arises in a beginner's mind while working in lab. Labs have different conditions according to respective areas so one section is devoted to each type of lab conditions. Considering incidences of minor and major accidents due to ignorance and negligence in laboratories, this manual has been prepared. This manual explains each and every topic in simple and easy language. It also includes do's and don'ts of lab so as to clearly envisage necessary instructions to be followed in labs.

The basic aim of this book is to provide comprehensive, rigorous, and balanced knowledge regarding lab conditions, potential threats and measures to be taken. It is a guide to understand lab environment. In writing this book, we had in mind its usefulness for students, researchers and lab workers. Authors' main focus was on to the point and brief discussion of the topics and hence made a successful handbook of the title.

Dr. A.K. Gupta
Pritanshu

Contents

Chapter 1 -Introduction

How safe are our laboratories? How safe are we? Are we aware of all possible laboratory accidents and most importantly do we know what to do in such mishappenings? A study conducted by University of California, Los Angeles concludes that scientists may have a false sense of security about the safety of their laboratories. More specific questions in the survey reveal that safety standards are often not adhered to.

Working in labs involves use of instrumentation, materials and reagents that have the potential to harm the handler, co-workers and perhaps even the environment. Consequently, it is momentous that principal investigators, employees, and students working in research labs should be equipped with professional knowledge and skills to safely and efficiently handle any emergency situations such as spills of hazardous materials, fires, explosions etc. For the event of an emergency, there should be an emergency response plan before hand to control the situation to insure everyone's good health and safety. Key elements of an emergency procedure plan are summarized by the acronym NEAR; Notify, Evacuate, Assemble, Report.

To avoid any kind of accident preventive measures should be followed by everyone working in lab. The following are some useful guidelines that applicable no matter what kind of project you are engaged in.

- Never engage in horseplay or rough housing in the laboratory.
- Do not bring any food or drink into the laboratory and do not eat, drink, or smoke there.
- Do not smell or taste any chemicals or other lab samples for any reason.
- Never work alone or unsupervised.
- Do not work when you are exhausted or emotionally upset.
- Don't perform any experiments that you haven't discussed in advance with your research advisor.
- Never leave experiments running unattended in the laboratory.
- Dress appropriately. This means your torso and arms should be well covered. Do not wear loose or sloppy clothing that could get caught in any equipment or come in contact with any chemicals. Long hair should be pulled back out of the way of any reagents or machinery.
- Wear the appropriate gloves, safety glasses or goggles, and a clean lab coat when handling chemical and/or biohazardous materials.
- Remove your gloves and wash your hands before using the keyboard associated with any instrumentation in the laboratory.

- Never pipette by mouth.
- Clean and disinfect all glassware, instrumentation, and lab surfaces after each experiment. Don't let sinks become filled with dirty glassware.
- Transport solvents and other reagents in secondary containers.
- If an accident or spill happens, be sure to notify your supervisor so that the appropriate protocols can be observed.
- Wash your hands before you exit the laboratory and especially after handling biohazards or chemical reagents.

Protective Personal Equipment

Personal protective equipment is a general term used to describe anything you can wear and/or use in order to protect yourself when working with chemical or biological hazards. Common examples of personal protective equipment include: footwear, lab coats, gloves, safety goggles and glasses, face shields, hard hats, respirators, and fume hoods.

Footwear

Closed toe, leather shoes provide the best general protection. Sandals, sneakers, etc. do not provide adequate protection in case of spills (biological or chemical hazards), or when handling heavy objects, tools, or involved in activities where heavy objects might fall onto the feet. If you will be involved with heavy machinery, steel-reinforced safety shoes may be required. There is also safety shoes specially designed to provide protection against extreme temperatures, caustic chemicals, and/or electrical hazards.

Lab Coat

Lab coats are normally worn in the research laboratory to protect your normal clothing against biological or chemical spills and to provide some additional body protection beyond that provided by your normal clothing. Important considerations in selecting an appropriate lab coat are the types of hazards (biological, chemical, fire, cold, etc.) to which you may be exposed. To be effective, the fabric should be resistant to the materials you are using. In addition, a lab coat should fit properly (you should be able to move comfortably in it with the coat buttoned or snapped down the front), be clean, and have long sleeves. Lab coats are normally provided by one's laboratory for the duration of the project.

Gloves

When handling chemical, physical, and/or biological hazards that can enter the body through the skin, it is important to wear the proper protective gloves. Note that there is no perfect glove: There is no kind of glove that will protect you from all hazards. There are several different kinds of gloves: disposable, fabric, leather, and metal mesh.

- **Disposable** - These are generally used to provide protection against biological or chemical hazards. There are two common kinds of gloves you will find in most biological and/or chemical laboratories - latex and nitrile gloves. Latex gloves provide good general protection in a biological research lab but provide no protection against common chemical hazards. Their use has decreased somewhat in recent years as some individuals have exhibited serious even life-threatening allergic reactions to latex. Nitrile gloves provide good general protection against a wide range of common solvents and chemical reagents. There are many other glove materials available which provide protection against particular chemical hazards. It is important to note that all glove materials are eventually permeated by some chemical reagent. Therefore to be maximally effective, the gloves should be changed whenever they become contaminated by the chemical reagent. The key to glove use is identifying the proper gloves for the job in question. For example, Kevlar gloves will provide good protection from extreme temperatures. Nitrile gloves provide good short term protection when handling a wide range of organic solvents and reagents. Depending on the nature of the hazards peculiar to your research project, you may find that you need several different kinds of gloves in order to be adequately protected. Best Glove Company's website is a good resource to consult when selecting the appropriate gloves for your work.
- **Fabric** - Cotton gloves are often used in pilot plants to absorb moisture and provide a better grip when working with heavy machinery.
- **Leather** - Leather gloves provide good protection when working with flames or when sparks may be present. They are also often worn together with insulated liners when working with electrical hazards.
- **Metal Mesh** - Metal mesh gloves are preferred when working with heavy machinery and/or cutting tools.

In addition to identifying the correct kind of glove, it is also important to make sure that the gloves that you use fit properly. Most gloves are commercially available in several different sizes. If you will be wearing gloves for an extended period of time (several hours or more), you may find it useful to purchase a box of disposable cotton glove liners that you can wear underneath your disposable gloves. Glove liners absorb perspiration and help minimize skin irritation.

Safety Glasses and Goggles

As a general rule, safety glasses with side shields should be worn at all times in the research laboratory even if you wear prescription glasses. Safety goggles rather than safety glasses are preferred whenever a chemical splash is a potential hazard. The side shields on safety glasses are simply not as effective as goggles in protecting your eyes from small particles and liquid splatter.

Most laboratories provide safety glasses or goggles to their researchers. If you wear contact lens underneath safety glasses be sure to consider the additional potential risk that your contact lenses may present if dust, caustic reagents or solvents get underneath your lenses and in your eyes. Removing your contact lenses in such a situation may take added time and increase your risk of injury.

Face Shields

A face shield should be worn whenever there the entire face needs protection. This means any time there is a potential that an aerosol of chemical or biological hazardous material may be created or whenever chemical or biohazards could splatter, or whenever there is the potential for flying particles or sparks (e.g., high pressure work, welding, soldering, machining, fire, explosion, etc.). A face shield should always be worn whenever handling tissue samples or animals where there is the potential for infectious transmission. Safety glasses or goggles should always be worn underneath a face shield for maximal protection.

Respirators

Respirators filter contaminants, either small airborne particles or chemicals including gases, out of the air. Whenever possible you should structure your work so that it can be carried out in a hood. If you are going to work with a respirator, be sure to obtain training prior to using this PPE. It is important to remember that to operate properly respirators must be regularly cleaned, sanitized (if biological hazards are involved), and maintained.

Chemical Fume Hoods

Whenever you use flammable or hazardous materials that pose an airborne or explosive hazard, you should work in a fume hood. Exposure is controlled in part through the moveable glass plate, the sash,that covers the front of the hood. Maximal protection is afforded when the sash, if it moves vertically, is closed or lowered as much as possible.

There are different types of fume hoods. Two of the most common types are the constant air volume (CAV) and the variable air volume (VAV) hoods. Constant air volume hoods are designed to maintain a constant air flow that doesn't vary when the hood sash is opened or closed. The disadvantage of these hoods is that the face velocity increases when the hood sash is lowered or decreases when the hood sash is rose as a result which can lead to either excessive turbulence or the escape of toxic materials from the hood. Variable air volume hoods are designed to maintain a constant face velocity whenever the hood sash is opened or closed minimizing air turbulence and maximizing user protection.

It is important that there be good airflow to the hood exhaust. Today most hoods are equipped with an airflow meter. These measure the face velocity which is the rate at which air is pulled into the hood exhaust.

Fume hoods should be inspected annually. Dated inspection stickers should be posted conspicuously somewhere on the front of the fume hood. All hoods are not the same. Depending on the hazards involved in your work, you may need to use a special kind of fume hood.

- Biosafety cabinets should be used when dealing with biological hazards.
- Chemical fume hoods should be used when flammable solvents and/or highly reactive reagents are involved.

Special fume hoods are required when working with certain radiological hazards such as iodine-125 or when working with perchlorates, which react explosively when mixed with organics. (http://www.louisville.edu/admin/dehs/lsfume.htm)

Prudent Practices

- Whenever possible endeavor to work with materials that are non-toxic or which present minimal health risks to you and your research group.
- Keep all materials at least 6 inches inside the fume hood. Doing this ensures you maximal protection in terms of hood air flow and air turbulence. A useful visual method of reminding yourself to do this is to place a strip of brightly colored labeling tape 6" lengthwise inside the hood.
- Never place beakers, pipettes, or other materials on the edge of the hood where they can be easily knocked off and where the hood provides no protective air flow.

- Keep the sash lowered at all times. When you are working in the hood, always keep the sash of the hood below your face.
- Regularly inspect the flow meter in your hood to ensure that the hood is functioning properly. If there isn't a flow meter contact your Office of Environmental Health and Safety. A simple, effective visual means of determining that there is hood air flow is to tape a Kimwipe at the bottom edge of the hood sash. If the hood is operating properly, it should be partially pulled inside by the hood's airflow.
- Do not put signs, or other materials that impede visual inspection of the hood's contents on the hood sash. In some organic and inorganic synthesis laboratories, it is common practice to write the chemical reaction on the face of the hood sash. If your lab does this be careful not to obscure your and others' direct view of the inside of the hood.
- Locate electrical devices outside the hood to avoid sparking that could ignite flammable reagents and/or solvents.
- Remember: hoods are not a substitute for good common sense. Do not do anything in a hood that you would not do on a desktop. For example, do not heat flammable solvents in an open beaker directly on a hot plate.
- Fume hoods should not be used to store hazardous materials. The bottles, glassware, and other materials that you place inside the fume hood can interfere with the proper airflow within the hood. Remove reagent bottles promptly when you are finished using them and replace them in their proper storage location in the laboratory.

CHAPTER 2 - GENERAL SAFETY

This chapter sets forth those practices which are deemed good safety practices common to all laboratory operations.

2.1 - GENERAL SAFETY AND OPERATIONAL RULES

A.   General Rules of Safety

1.   No running, jumping, or horseplay in laboratory.

2.   No employee should work alone in a laboratory or chemical storage area when performing a task that is considered unusually hazardous by the laboratory supervisor or safety officer.

3.   Spills should be cleaned immediately. Specifics of emergency spill tactics are provided in the Emergency Response chapter of this manual. Water spills can create a hazard because of the slip potential and flooding of instruments (particularly on the floor below.) Small spills of liquids and solids on bench tops shall be cleaned immediately to prevent contact with skin or clothing.

4.   Ladders should be in good condition and used in the manner for which they were designed. Wooden ladders shall not be covered with paint or other coating. (Structural defects may be hidden by the coating.)

5.   Lifting of heavy items must be performed in the proper fashion, using the legs to lift, and not the back.

6.   It is the responsibility of everyone working in the laboratory to make certain that the laboratory is left clean after work is performed.

7.   Outsiders, children and animals, except for those that are the subject of experimentation are to be excluded from all University laboratory areas.

B.   Personal Hygiene

1.   Wash promptly whenever a chemical has contacted the skin. Know what you are working with and have the necessary cleaning/neutralization material on hand and readily available.

2.   No sandals, open toed shoes or clogs should be worn by laboratory personnel.

3.   Clothing worn in the laboratory should offer protection from splashes and spills. Laboratory clothing should be kept clean and replaced when necessary. Lab coats are

not to be worn outside the laboratory, especially in rest room or break facilities. Any lab coats, respirators, or other protective gear must be left in the lab areas. Employees must, as a matter of routine, be responsible for washing, cleaning, and any other decontamination required when passing between the lab and the other areas. Washing should be done with soap and water; **do not** wash with solvents.

6.   **Never** pipette by mouth. **Always** use a bulb to pipette.

8.   Do not drink, eat, smoke, or apply cosmetics in the laboratory or chemical storage areas.

9.   Do not use ice from laboratory ice machines for beverages.

10.   No food, beverage, tobacco, or cosmetics products are allowed in the laboratory or chemical storage areas at any time. Cross contamination between these items and chemicals or samples is an obvious hazard and should be avoided.

C.   Housekeeping

As in many general safety procedures, the following listing of good housekeeping practices indicates common sense activities which should be implemented as a matter of course in the laboratory. These recommendations are designed for accident prevention.

1.   The area must be kept as clean as the work allows. Each laboratory employee must be responsible for maintaining the cleanliness of his/her area.

2.   Reagents and equipment items should be returned to their proper place after use. This also applies to samples in progress. Contaminated or dirty glassware should be placed in specific cleaning areas and not allowed to accumulate.

3.   Chemicals, especially liquids, should never be stored on the floor, except in closed door cabinets suitable for the material to be stored. Nor should large bottles (2.5l or larger) be stored above the bench top.

4.   Counter tops should be kept neat and clean. Bench tops and fume hoods shall not be used for chemical storage. Stored items or equipment shall not block access to the fire extinguisher(s), safety equipment, exits or other emergency items.

5.   All containers must be properly labeled with MSDS.

D.   Electrical

1.   All electrical equipment must be properly grounded.

2.   All electrical equipment shall be U.L. listed and/or F.M. approved.

3.   Sufficient room for work must be present in the area of breaker boxes. All the circuit breakers and the fuses shall be labeled to indicate whether they are in the "on" or "off" position, and what appliance or room area is served.. Fuses must be properly rated.

4.   Equipment, appliance and extension cords should be in good condition. Extension cords shall not be used as a substitute for permanent wiring.

6.   Electrical cords or other lines should not be suspended unsupported across rooms or passageways. Do not route cords over metal objects such as emergency showers, overhead pipes or frames, metal racks, etc. Do not run cords through holes in walls or ceilings or through doorways or windows. Do not place under carpet, rugs, or heavy objects. Do not place cords on pathways or other areas where repeated abuse can cause deterioration of insulation.

7.   Multi-outlet plugs should not be used unless they have a built-in circuit breaker. This causes overloading on electrical wiring, which will cause damage and possible overheating.

8.   All building electrical repairs, splices, and wiring should be performed by the trained personnel.

E.   Vacuum Operations

In an evacuated system, the higher pressure is on the outside, rather than the inside, so that a break causes an implosion rather than an explosion. The resulting hazards consist of flying glass, spattered chemicals, and possibly fire.

A moderate vacuum, such as 10 mm Hg, which can be achieved by a water aspirator, often seems safe compared with a high vacuum, such as 10-5 mm Hg. These numbers are deceptive, however, because the pressure differences between the outside and inside are comparable. Therefore any evacuated container must be regarded as an implosion hazard.

1.   When working with a vacuum be aware of implosion hazards. Apply vacuum only to glassware specifically designed for this purpose, i.e., heavy wall filter flasks, desiccators, etc.

2.   Never evacuate scratched, cracked, or etched glassware. Always check for stars or cracks before use.

3.   Vacuum glassware which has been cooled to liquid nitrogen temperature or below should be annealed prior to reuse under vacuum.

Rotary evaporator condensers, receiving flasks, and traps should be taped or kept behind safety shields when under a vacuum.

5.   All condensers connected to rotary evaporators should at least be cooled with circulating ice water.

6.   The use of a vacuum for the distillation of the more volatile solvents, e.g. ether, low boiling petroleum ether and components, methylene chloride, etc., should be avoided whenever possible. In situations requiring reduced pressure, two alternatives should be considered; 1) Utilization of Rotovac System, or 2) Solvent recovery via atmospheric pressure distillation (preferred method).

7.   Water, solvents, or corrosive gases should not be allowed to be drawn into a building vacuum system.

8.   When a vacuum is supplied by a compressor or vacuum pump to distill volatile solvents, a cold trap should be used to contain solvent vapors. Cold traps should be of sufficient size and low enough temperature to collect all condensable vapors present in a vacuum system. If such a trap is not used, the pump or compression exhaust must be vented to the outside using explosion proof methods.

9.   After completion of an operation in which a cold trap has been used, the system should be vented. This venting is important because volatile substances that have been collected in the trap may vaporize when the coolant has evaporated and cause a pressure buildup that could blow the apparatus apart.

10.   After vacuum distillations, the pot residue must be cooled to room temperature before air is admitted to the apparatus.

11.   All desiccators under vacuum should be completely enclosed in a shield or wrapped with friction tape in a grid pattern that leaves the contents visible and at the same time guards against flying glass should the vessel collapse. Various plastic (e.g., polycarbonate) desiccators now on the market reduce the implosion hazard and may be preferable.

SECTION 2.2 - GENERAL SAFETY EQUIPMENT

Workers in a laboratory environment are surrounded by physical and chemical hazards, and the potential for accident and injury is always present. Adequate safety equipment in good working order shall be provided to prevent accidents and injury.

Fire Extinguishers

Safety Showers.

Eyewash Fountains.

First Aid Kits.

Biological Saftey Cabinets.

Ventilation Hoods.

Safety Shields.

Section 2.3 - Compressed Gas Safety

Compressed gases can be combustible, explosive, corrosive, poisonous, inert, or a combination of hazards.

1.   Placed properly to prevent tipping.

2.   New cylinders should be inspected for leakage and clear labels.

1. Identification should be stenciled or stamped on the cylinder or a label for easy, right and complete determination of content.

2.   All gas lines leading from a compressed gas supply should be clearly labeled to identify the gas, the laboratory served, and the relevant emergency telephone numbers. The labels should be color coded to distinguish hazardous gases (such as flammable, toxic, or corrosive substances) (e.g., a yellow background and black letters). Signs should be conspicuously posted in areas where flammable compressed gases are stored, identifying the substances and appropriate precautions (e.g., HYDROGEN - FLAMMABLE GAS - NO SMOKING - NO OPEN FLAMES).

3.   All cylinders containing flammable gases should be stored in a well-ventilated area and away from fire sources.

4.   Oxygen cylinders, full or empty, should be stored away from flammable gases and greasy and oily materials.

5.   The main cylinder valve should be closed as soon as it is no longer necessary that it be open (i.e., it should never be left open when the equipment is unattended or not operating).

6.   Cylinder valves should be opened slowly. Main cylinder valves should never be opened all the way.

7.   When opening the valve on a cylinder containing an irritating or toxic gas, the user should position the cylinder with the valve pointing away from them and warn those working nearby.

8   Regulators are gas specific and not necessarily interchangeable.

9.   Piping material must be compatible with the gas being supplied.

10.   A cylinder should never be emptied to a pressure lower than 172 kpa (25 psi/in2) (the residual contents may become contaminated if the valve is left open). When work involving a compressed gas is completed, the cylinder must be turned off, and if possible, the lines bled. When the cylinder needs to be removed or is empty (see above), all valves shall be closed, the system bled, and the regulator removed. The valve cap shall be replaced, the cylinder clearly marked as "empty," and returned to a storage area for pickup by the supplier. Empty and full cylinders should be stored in separate areas.

11.   Where the possibility of flow reversal exists, the cylinder discharge lines should be equipped with approved check valves to prevent inadvertent contamination of cylinders connected to a closed system. "Sucking back" is particularly troublesome where gases are used as reactants in a closed system. A cylinder in such a system should be shut off and removed from the system when the pressure remaining in the cylinder is at least 172 kpa (25 psi/in2). if there is a possibility that the container has been contaminated, it should be so labeled and returned to the supplier.

12.   Liquid bulk cylinders, if used, should be clearly marked for each valve and must be placed in well ventilated area.

13.   Always use safety glasses (preferably a face shield) when handling and using compressed gases.

14.   During transportation cylinders must not be subjected to rough handling or abuse
.

15. To protect the valve during transportation, the cover cap should be screwed on hand tight and remain on until the cylinder is in place and ready for use.

16. Cylinders should never be rolled or dragged.

17. Only one cylinder should be handled (moved) at a time.

Cryogenic liquids

Neither liquid nitrogen nor liquid air should be used to cool a flammable mixture in the presence of air because oxygen can condense from the air and lead to a potentially explosive condition.

2. Adequate ventilation must always be used to prevent the build-up of vapors of flammable gases such as hydrogen, methane, and acetylene.

3. Adequate ventilation is also required when using gases such as nitrogen, helium, or hydrogen. In these cases, oxygen can be condensed out of the atmosphere creating a potential for explosive conditions.

4. Cylinders and other pressure vessels used for the storage and handling of liquefied gases should not be filled to more than 80% of capacity, to prevent the possibility of thermal expansion and the resulting bursting of the vessel by hydrostatic pressure.

5. Appropriate impact-resistant containers must be used that have been designed to withstand the extremely low temperatures.

6. Even very brief contact with a cryogenic liquid is capable of causing tissue damage similar to that of thermal burns. Prolonged contact may result in blood clots that have potentially serious consequences. In addition, surfaces cooled by cryogenic liquids can cause severe damage to the skin. Gloves and eye protection (preferably a face shield) should be worn at all times when handling cryogenic liquids. Gloves should be chosen that are impervious to the fluid being handled and loose enough to be tossed off easily. Appropriate dry gloves should be used when handling dry ice. "Chunks" or cubes should be added slowly to any liquid portion of the cooling bath to avoid foaming over.

7. As the liquid form of gases warm and become airborne, oxygen may be displaced to the point that employees may experience oxygen deficiency or asphyxiation. Any area where such materials are used should be well ventilated.

For this same reason, employees should avoid lowering their heads into a dry ice chest. (Carbon dioxide is heavier than air, and suffocation can result.)

SECTION 2.4- SAFETY PRACTICES FOR DISPOSAL OF BROKEN GLASSWARE

Inspect all glassware before use. Do not use broken, chipped, starred or badly scratched glassware.

1. Uncontaminated glassware can be dispose in a rigid, puncture proof container while contaminated glassware must be disinfected before disposal and container must be labeled.

2. For radioactive glassware follow specific instructions for radioactive materials.

3.   Chemical container should be emptied before disposal if safe to do so.

4. Any doubt (infection, hazard) regarding contents of glassware must be clarified before disposal to prevent any kind of accident.

SECTION 2.5 - CENTRIFUGE SAFETY

1. Always review the owner's manual for type of centrifuge, rotar, tube size and adapter, speed and length of time.
2. Insure the tube fits properly in the rotor.
3. Insure you are using the appropriate level of containment. If the material potentially infectious/radioactive use aerosol containment tubes and perform loading and unloading in a biological safety cabinet.
4. Insure centrifuge bowl and tubes are dry.
5. Is the centrifuge spindle clean?
6. Insure rotor is properly seated on drive hub.
7. Make sure tubes are properly balanced in rotor (½ gram at 1 G is roughly equivalent to 250 Kg @ 500,000 G's).
8. Are O-rings properly attached to the rotor? Is the vacuum grease fresh?
9. Has the rotor been properly secured to drive?
10. Is the centrifuge lid shut properly?
11. Make sure the run is proceeding normally before you leave the area.

12. Make sure the rotor has STOPPED completely before you open the centrifuge lid; then check for spills. If it is there follow proper decontamination and cleaning steps.

Maintenance/Cleaning:

1. Keep rotors clean and dry. If spills occur, make sure rotor has been cleaned/ decontaminated. If salts or corrosive materials were used, ensure they have been removed from the rotor.

2. Avoid mechanical scratches. The smallest, scarcely visible scratch allows etching to enlarge the fracture, which is subject to enormous rupturing forces at high G's--a vicious cycle leading to rotor explosion.

3. Avoid bottle brushes with sharp metal ends and harsh detergents when cleaning aluminum rotor heads.

4. After proper clean-up, rinse the rotor with de-ionized water.

Inspections:

1. Check the rotor for rough spots, pitting, and discoloration. If discovered, check with the manufacturer before using. Use professional rotor inspection services frequently. These visits can be arranged to accommodate numerous users throughout the University.

2. Consult the centrifuge manufacturer and centrifuge log for the derating schedule for the rotor. Remember--an unlogged ultra-speed centrifuge is a ticking time bomb.

UNIVERSITY RESPONSIBILITIES

1. Keep records of employee exposures to hazardous chemicals. Records should include measurements made to monitor exposures, if any, as well as any medical consultations and examinations, including written opinions.

2. Provide university employees with training and information regarding chemical and physical hazards. Also, access to medical consultation and examinations.

3. For incoming hazardous chemicals:

a. Require that the incoming hazardous chemicals have adequate labels. Do not allow the removal or defacement of these labels.

b. Require that the msds for incoming hazardous chemicals be on hand prior to receipt of hazardous chemicals whenever possible. Require that msds be acquired for all hazardous chemicals on hand whenever possible.

c. keep all material safety data sheets (msds) that the university receives.

d. Make msds accessible to employees.

e. Maintain an accurate inventory of all chemicals in university laboratories.

4. When hazardous chemicals are generated in university laboratories:

a. If the hazardous properties are known, train university employees.

b. If the hazardous properties are not known, treat the chemical as though it is hazardous and provide protection.

5. If selected carcinogens, reproductive toxins, or acute toxins that are very highly toxic are used in the laboratory, identify and post one or more areas as "designated area(s)."

Individual Responsibilities

Responsibility for chemical hygiene rests at all levels including the:

1. *University President*, who has ultimate responsibility for chemical hygiene within University and must, with other administrators, provide continuing support for University chemical hygiene.

2. *Supervisor of a College, Department or other administrative unit*, who is responsible for chemical hygiene in that unit.

3. *Departmental Chemical Hygiene Officers (If appointed)*, who have overall responsibility for chemical hygiene in all departmental laboratories.

4. *Project director or director of other specific operation*, who has primary responsibility for chemical hygiene procedures for that operation, and is responsible for:

a. Maintaining an accurate Laboratory Chemical Inventory List. Insure that the Departmental Chemical Hygiene Officer receives copies of this list as necessary. (See University Policy and Procedures Letter 3-0535 "Hazard Communications Program."

b. Ensuring that workers know and follow the chemical hygiene rules,

c. Ensuring that protective equipment is available and in working odor,

d. Ensuring that all containers in the work area are properly labeled,

e. Ensuring that MSDS's are maintained for each hazardous substance in the laboratory and ensuring that they are readily accessible to laboratory employees,

f. Ensuring that appropriate training has been provided to all employees,

g. Providing regular, formal chemical hygiene and housekeeping inspections including routine inspections of emergency equipment,

h. Knowing the current legal requirements concerning regulated substances,

i. Determining the required levels of protective apparel and equipment, and

j. Ensuring that facilities for use of any material being ordered are adequate.

5. *Laboratory worker*, who is responsible for:

a. Planning and conducting each operation in accordance with safe procedures; and

b. Developing and maintaining good personal chemical hygiene habits.

The Content of the Chemical Hygiene Plan

The chemical hygiene plan should include each of the following elements and shall also indicate the specific measures to be taken to ensure that University employees are protected.

1. Standard operating procedures relevant to all laboratory operations, to be followed by laboratory employees.

2. Statements of the criteria that will be used to determine and implement control measures to reduce employee exposure to hazardous chemicals. These measures include engineering controls, use of personal protective equipment,

and personal hygiene practices. Criteria to reduce exposure to extremely hazardous chemicals used in the laboratory shall be specifically included.

3. A requirement that fume hoods and other protective equipment shall function properly and descriptions of the methods to be taken to make sure that such equipment is functioning properly.

4. Provisions for employee training and information.

5. Circumstances under which a laboratory practice requires prior approval from a supervisor before implementation.

6. Provisions for medical consultation and examination.

7. Designation of personnel responsible for implementation of the chemical hygiene plan.

8. Provisions for additional protection for employees when working with particularly hazardous substances, including:

a. *Select carcinogens.*

b. *Reproductive toxins.*

c. *Substances with a high degree of acute toxicity.*

9. Specific mention of the following provisions, including when appropriate:

a. *Establishment of a designated area.*

b. *Use of containment devices such as fume hoods or glove boxes.*

c. *Procedures for safe removal and disposal of contaminated and hazardous waste; and*

d. *Decontamination procedures.*

Exposure Assessments, Medical Consultations, and Examinations

1. Suspected Exposures to Toxic Substances

There may be times when employees or supervisors suspect that an employee has been exposed to a hazardous chemical to a degree and in a manner that might have caused harm to the victim. If the circumstances suggest a reasonable suspicion of exposure, the victim is entitled to a medical consultation and, if so determined in the consultation, also to a medical examination. All medical examinations and consultations shall be provided without cost to the employee, without loss of pay, and at a reasonable time and place.

Criteria for Reasonable Suspicion of Exposure

All exposure complaints and their disposition, no matter what the ultimate disposition may be, are to be documented by the respective department chemical hygiene officer using the employer's first notice of injury form and employee exposure report form. Copies of these forms shall be sent to university personnel services. If no further assessment of the event is deemed necessary, the reason for that decision shall be included on the employee exposure report form. If the decision is to investigate, a formal exposure assessment will be initiated by the departmental chemical hygiene officer. Environmental health services and university personnel services office of risk management shall provide the formal exposure assessment.

1. *EXPOSURE ASSESSMENT .IN CASES OF EMERGENCY, EXPOSURE ASSESSMENTS ARE CONDUCTED AFTER THE VICTIM HAS BEEN TREATED, OTHERWISE EXPOSURE ASSESSMENTS SHOULD BE COMPLETED BEFORE MEDICAL CONSULTATIONS ARE UNDERTAKEN.*

NOTE: IT IS NOT THE PURPOSE OF AN EXPOSURE ASSESSMENT TO DETERMINE THAT A FAILURE ON THE PART OF THE VICTIM, OR OTHERS, TO FOLLOW PROPER PROCEDURES WAS THE CAUSE OF AN EXPOSURE. THE PURPOSE OF AN EXPOSURE ASSESSMENT IS TO DETERMINE THAT THERE WAS, OR WAS NOT, AN EXPOSURE THAT MIGHT HAVE CAUSED HARM TO ONE OR MORE EMPLOYEES AND, IF SO, TO IDENTIFY THE HAZARDOUS CHEMICAL OR CHEMICALS INVOLVED. OTHER INVESTIGATIONS MIGHT WELL USE RESULTS AND CONCLUSIONS FROM AN EXPOSURE ASSESSMENT, ALONG WITH OTHER INFORMATION, TO DERIVE RECOMMENDATIONS THAT WILL PREVENT OR MITIGATE ANY FUTURE OVEREXPOSURES. HOWEVER, EXPOSURE ASSESSMENTS DETERMINE FACTS; THEY DO NOT MAKE RECOMMENDATIONS.

(a)   Unless circumstances suggest other or additional steps, these actions constitute an exposure assessment:

i.   *Interview the complainant and also the victim, if not the same person.*

ii.   *List the essential information about the circumstances of the complaint, including:*

-     *The chemical under suspicion.*

-     *Other chemicals used by victim.*

-     *All chemicals being used by others in the immediate area.*

-      *Other chemicals stored in that area.*

-     *Symptoms exhibited or claimed by the victim.*

-     *How these symptoms compare to symptoms stated in the materials safety data sheets for each of the identified chemicals.*

-     *Were control measures, such as personal protective equipment and hoods, used properly?*

-     *Were any air sampling or monitoring devices in place?   If so, are the measurements obtained from these devices consistent with other information?*

(b)   Monitor or sample the air in the area for suspect chemicals.

(c)   Determine whether the victim's symptoms compare to the symptoms described in the MSDS or other pertinent scientific literature.

 (2)   Notification of Results of Monitoring.

Within fixed and limited period of time receipt of the results of any monitoring, notify affected employees of those results.

<u>Medical Consultation and Examination</u>

- ✓ If employees feel that they have been exposed to hazardous chemicals, employees are required to arrange for an Exposure Assessment. The Exposure Assessment will be utilized by the consulting physician to determine if further medical consultations and examinations are warranted.

The purpose of a medical consultation is to determine whether a medical examination is warranted. When, from the results of an Exposure Assessment, it is suspected or known that an employee was overexposed to a hazardous chemical or chemicals, the employee should obtain medical consultation from or under the direct supervision of a licensed physician.

When warranted, employees also should receive a medical examination from or under the direct supervision of a licensed physician who is experienced in treating victims of chemical overexposure. The medical professional should also be knowledgeable about which tests or procedures are appropriate to determine if there has been an overexposure; these diagnostic techniques are called "differential diagnoses."

Provide the physician with:

(1)   The identity of the hazardous chemicals to which the employee may have been exposed.

(2)   The exposure conditions.

(3)   The signs and symptoms of exposure the victim is experiencing, if any.

A statement that the employee has been informed both of the results of the consultation or examination and of any medical condition that may require further examination or treatment.

These written statements and records should not reveal specific findings that are not related to an occupational exposure. All memos, notes, and reports related to a complaint of actual or possible exposure to hazardous chemicals are to be maintained as part of the record. Employees should be notified of the results of any medical consultation or examination with regard to any medical condition that exists or might exist as a result of overexposure to a hazardous chemical. And records must be kept regarding whole incidence.

Chapter 3-Biological Safety

Biological contamination can pose severe threat to working personnel and others. It is therefore becomes necessary to follow every precaution and rules to avoid any chance of contamination. There are labs where personnel work with hazardous viruses and bacteria. These are nothing but noxious to your health. Personnel have contracted infections in the laboratory throughout the history of microbiological and biohazard research. A number of cases have been attributed to carelessness or poor technique in the handling of infectious materials.

IN ANY CASE NO LABORATORY RULE MUST BE AVOIDED.

The three elements laboratory practice and technique, safety equipment, and facility design play role to reduce exposure of laboratory workers and other persons, and to prevent escape into the outside environment of potentially hazardous agents.

3.1 Laboratory Practice and Technique

The most important element is strict adherence to standard microbiological practices and techniques. Persons working with infectious agents or infected materials must be aware of potential hazards and must be trained and proficient in the practices and techniques required for safely handling such material. The director or person in charge of the laboratory is responsible for providing or arranging for appropriate training of personnel.

When standard laboratory practices are not sufficient to control the hazard associated with a particular agent or laboratory procedure, additional measures may be needed. The laboratory supervisor is responsible for selecting additional safety practices, which must be in keeping with the hazard associated with the agent or procedure.

Laboratory personnel, safety practices and techniques must be supplemented by appropriate facility design and engineering features, safety equipment and management practices.

1. Engineering controls should be examined and maintained or replaced on a regular schedule to ensure their effectiveness.

2. Employees must wash their hands immediately or as soon as possible after removal of gloves or other personal protective equipment and after hand contact with blood or other potentially infectious materials.

3.   All personal protective equipment should be removed immediately upon leaving the work area or as soon as possible if overtly contaminated and placed in an appropriately designated area or container for storage, washing, decontamination or disposal.

4.   Used needles and other sharps should not be sheared, bent, broken, recapped, or resheathed by hand. Used needles should not be removed from disposable syringes.

5.   Eating, drinking, smoking, applying cosmetics or lip balm, and handling contact lenses are prohibited in work areas where there is a potential for occupational exposure.

6.   Food and drink should not be stored in refrigerators, freezers, or cabinets where blood or other potentially infectious materials are stored or in other areas of possible contamination.

7.   All procedures involving blood or other potentially infectious materials should be performed in such a manner as to minimize splashing, spraying, and aerosolization of these substances.

3.2   Safety Equipment (Primary Barriers)

Safety equipment includes biological safety cabinets and a variety of enclosed containers. The biological safety cabinet is the principal device used to provide containment of infectious aerosols generated by many laboratory procedures. Open fronted Class I and Class II biological safety cabinets are partial containment cabinets which offer significant levels of protection to laboratory personnel and the environment when used with good microbiological techniques. The gas-tight Class III biological safety cabinet provides the highest attainable level of protection to personnel and the environment. Further references on proper and effective use of biological safety cabinets may be found in "Effective Use of Biological Safety Cabinets."

An example of an enclosed container is the safety centrifuge cup, which is designed to prevent aerosols from being released during centrifugation.

3.3   Personal Protective Equipment

When there is a potential for occupational exposure appropriate personal protective equipment such as gloves, gowns, fluid-proof aprons, laboratory coats, head and foot coverings, face shields or masks, eye protection, mouthpieces, resuscitation bags, pocket masks, or other ventilation devices must be used.

These personal protective devices are often used in combination with biological safety cabinets and other devices which contain the agents, animals, or materials being examined. In some situations in which it is impractical to work in biological safety cabinets, personal protective devices may form the primary barrier between personnel and the infectious materials. Examples of such activities include certain animal studies, animal necropsy, production activities, and activities relating to maintenance, service or support of the laboratory facility. Gloves must be worn when the employee has the potential for the hands to have the direct skin contact with blood, other potentially infectious materials, mucous membranes, non-intact skin, and when handling items or surfaces soiled with blood or other potentially infectious material.

    a.   Disposable (single-use) gloves such as surgical or examination gloves shall be replaced as soon as possible when visibly soiled, torn, punctured or when their ability to function as a barrier is compromised. They shall not be washed or disinfected for re-use.

    b.   Utility gloves may be disinfected for re-use if the integrity of the glove is not compromised, however, they must be discarded if they are cracked, peeling, discolored, torn, punctured, or exhibit other signs of deterioration.

5. Masks and eye protection or chin-length face shields must be worn whenever splashes, spray, spatter, droplets, or aerosols of blood or other potentially infectious materials may be generated and there is a potential for eye, nose, or mouth contamination. Appropriate protective clothing should be worn when the employee has potential for occupational exposure. The type and characteristics will depend upon the task and degree of exposure anticipated.

    a.   Gowns, lab coats, aprons or similar clothing shall be worn if there is a potential for soiling of clothes with blood or other potentially infectious materials.

    b.   Fluid resistant clothing, surgical caps or hoods shall be worn if there is a potential for splashing or spraying of potentially infectious materials.

    c.   Fluid-proof shoe covers shall be worn if there is a potential for shoes to become contaminated and/or soaked potentially infectious materials.

3.4 Housekeeping

1. WORK SURFACES SHOULD BE DECONTAMINATED WITH AN APPROPRIATE DISINFECTANT AFTER COMPLETION OF PROCEDURES

2. Protective coverings such as plastic wrap, aluminum foil, or imperviously-backed absorbent paper may be used to cover equipment and environmental surfaces. These coverings should be removed and replaced at the end of the work shift or when they become overtly contaminated.

3. Equipment which may become contaminated with blood or other potentially infectious materials shall be checked routinely and disinfected.

4. All bins, pails, cans, and similar receptacles intended for re-use which have a potential for becoming contaminated with potentially infectious materials shall be inspected, cleaned, and disinfected on a regularly scheduled basis

5. Broken glassware which may be contaminated should not be picked up directly with the hands. It should be cleaned up using mechanical means such as a brush and dust pan, tongs, cotton swabs or forceps.

6. Reusable items contaminated with potentially infectious materials shall be decontaminated prior to washing and/or reprocessing.

3.5 Infectious Waste Disposal

All infectious waste destined for disposal shall be placed in closable, leak-proof containers or bags that are color-coded or labeled.

3.6 Bio-safety Levels

Four bio-safety levels are described which consist of combinations of laboratory practices and techniques, safety equipment, and laboratory facilities appropriate for the operations performed and the hazard posed by the infectious agents and for the laboratory function or activity.

1.   Bio-safety Level 1: Practices, safety equipment, and facilities are appropriate for facilities in which work is done with defined and characterized strains of viable microorganisms not known to cause disease in healthy adult humans. *Bacillus subtilis*, *Naegleria gruberi*, and infectious canine hepatitis virus are representative of those microorganisms meeting these criteria. Many agents not ordinarily associated with disease processes in humans are, however, opportunistic pathogens and may cause infection in the young, the aged, immunodeficient or immunosuppressed individuals. Vaccine strains which have undergone multiple in-vivo passages should not be considered avirulent simply because they are vaccine strains.

2.   Bio-safety Level 2: Practices, equipment, and facilities are applicable to clinical facilities in which work is done with the broad spectrum of indigenous moderate-risk agents present in the community and associated with human disease of varying severity. With good microbiological techniques, these agents can be used safely in activities conducted on the open bench, provided the potential for producing aerosols is low. Hepatitis B virus, the *Salmonellae*, and *Toxoplasma* spp. are representative of microorganism assignment to this containment level. Primary hazards to personnel working with these agents may include accidental autoinnoculation, ingestion, and skin or mucous membrane exposure to infectious materials. Procedures with high aerosol potential that may increase the risk of exposure to personnel, must be conducted in primary containment equipment or devices.

3.   Bio-safety Level 3: Practices, safety equipment, and facilities are applicable to facilities in which work is done with indigenous or exotic agents where the potential for infection by aerosols is real and the disease may have serious or lethal consequences. Autoinnoculation and ingestion also represent primary hazards to personnel working with these agents. Examples of such agents for which bio-safety Level 3 safeguards are generally recommended include *Mycobacterium tuberculosis*, St. Louis encephalitis virus and *Coxiella burnetti*.

4.   Bio-safety Level 4: Practices, safety equipment, and facilities are applicable to work with dangerous and exotic agents which pose a high individual risk of life threatening disease. All manipulations of potentially infectious diagnostic materials, isolates, and naturally or experimentally infected animals pose a high risk of exposure and infection to laboratory personnel. Lassa Fever virus is representative of the microorganisms assigned to Level 4.

It is the responsibility of the laboratory supervisor to establish standard procedures in the laboratory which realistically address the issue of ineffective hazard of clinical specimens.

3.7 - EFFECTIVE USE OF BIOLOGICAL SAFETY CABINETS

Technology has provided safety equipments but these can fulfill their role only if personnel complement them. It means a worker must be fully trained and devoted to avoid mistakes or human errors

The Laminar Flow Biological Safety Cabinet, designed to prevent escape of pathogens into the workers' environment and to bar contaminants from the research work zone, is a key element to safe, successful experimentation with biological materials. Escape of pathogens into the workers' area is prevented by an air barrier at the front opening and the cleaning action of the exhaust air filter. Inward flow of room air into the front air intake grill creates the air barrier. The amount of air drawn into the air intake grill and the amount of air exhausted through the exhaust filter are equal. The exhaust filter removes airborne biological contaminants which may be released in the cabinet. It does not remove chemical or radiological contaminants.

Contamination of the work area inside the cabinet is prevented by the cleaning action of the supply filters. Air flows through the cabinet work area in a downward direction at a uniform velocity. The air continues to be recirculated by the fan through the air flow plenum. Airborne biological contaminants are removed by the filters as the air is returned to the cabinet work area.

There are three start-up steps:

1.   Turn on the lights

2.   Check the air intake and exhaust grill to make sure they are unobstructed

3.   Turn on the fan. Allow the fan to operate a minimum of five minutes before manipulations are begun in the cabinet.

1.   Some cabinets are equipped with ultraviolet light. These must be turned off during the day while laboratory personnel are occupying the room.

2.   Hands and arms should be washed well with germicidal soap before and after work in the cabinet.

3.   Technicians are encouraged to wear long-sleeve gowns with knit cuffs and rubber gloves. This minimizes the shedding of skin flora into the work area and protects the hands and arms from contamination by viable agents.

4. Interior surfaces of the work area should be disinfected by wiping them thoroughly with 70% alcohol.

5. The cabinets should not be overloaded. Everything needed for the complete procedure should be placed in the cabinet before starting so that nothing passes in or out through the air barrier until the procedure is completed.

6. Do not place anything over the front intake or rear exhaust grill in units having a solid work surface. As a general rule, keep equipment at least four inches inside the cabinet window and perform transfer of viable materials as deeply into the cabinet as possible.

7. After all materials have been placed in the cabinet, wait 2-3 minutes before beginning work. This will allow sufficient time for the cabinet air to purge airborne contamination from the work area.

8. Hold the activity in the room to a minimum. Unnecessary activity may create disruptive air currents. The ideal location for a cabinet is in a quiet end of the laboratory, removed from doorways, air conditioning and heating vents. Opening and closing laboratory doors can cause disruptive drafts that allow microorganisms to penetrate the air barrier.

9. Schedule uninterrupted work periods. The movement of objects including hands and arms causes turbulent air currents which disrupt the air barrier and allow escape and entrance of airborne contaminants.

10. Air turbulence caused by rotating laboratory equipment, such as a small clinical centrifuge, disrupt air flow within the cabinet and at the work opening. This is sufficient for contaminated air to escape to the laboratory environment. If a centrifuge must be used in the cabinet, do not perform other research activities in the cabinet while the centrifuge is operating.

11. Normal laboratory contamination control procedures and aseptic techniques are still necessary while working in the biological safety cabinet.

12. Equipment in direct contact with the biological agent should not be removed from the cabinet until enclosed or until the surface is decontaminated. Trays of discarded pipettes and glassware must be covered before removal from the cabinets.

13. If an accident occurs which spills or splatters the biological agent in the work area, all surfaces in the cabinet must be surface decontaminated before being removed.

15.   Do not mouth pipette.

16.   Following completion of the work, the following steps must be performed:

A.   Allow the cabinet to run 2-3 minutes with no activity. This will allow sufficient time for cabinet air flow to purge airborne contaminants from the work area;

B.   Decontamination of the interior surfaces should be repeated after removal of all materials, cultures, apparatus, etc. A careful check of the work area should be made for spilled or splashed nutrients. They may support fungus growth and result in spore liberation that contaminates the protected work environment; and

C.   Shut down by turning off the fan and lights. Use UV lights according to manufacturer's recommendations. Do not use the cabinet to store excess laboratory equipment.

SECTION 3.8 - BIOHAZARD WASTE

A.   Biohazard Wastes are discarded materials "that are biological agents or conditions (as an infectious organism or unsecure laboratory condition) that constitutes a hazard to man or his environment." This definition includes "any and all substances which contain materials to which organisms may cause injury or disease to man or his environment, but which are not regulated as controlled industrial waste".

B.   Infectious Wastes include the following categories:

- cultures and stocks of infectious agents and associated biologicals;

- human blood and blood products,

- pathological wastes,

- contaminated sharps,

- contaminated animal carcasses, body parts, and bedding,

- wastes from surgery, necropsy and other medical procedures,

- laboratory wastes,

- isolation wastes, unless determined to be non-infectious by the infection control committee at the health care facility,

- any other material and contaminated equipment which, in the determination of the facility infection control staff, presents a significant danger of infection because it is contaminated with, or may reasonably be expected to be contaminated with, etiologic agents.

C.   Chemical Wastes subject to the requirements of biohazard waste regulations include wastes from the following categories:

- pharmaceutical wastes,

- laboratory reagents contaminated with infectious body fluids,

- all the disposable materials which have come into contact with cytotoxic/antineoplastic   agents during the preparation, handling, and administration of such agents, and

- other chemicals that may be contaminated by infectious agents, as designated by experts at the point of generation of the waste.

D.   Treated Biohazard Wastes are all biohazard wastes that have been treated by one of the following methods and rendered harmless and biologically inert:

- incineration in an approved incinerator,

- steam sterilization at sufficient time and temperature to destroy infectious agents in waste ("autoclaved"),

- chemical disinfection where contact time, concentration, and quantity of the chemical disinfectant are sufficient to destroy infectious agents in the waste, and

- any other method generally recognized as effective.

E.   Sharps are used in animal or human patient care or treatment or in medical research, or industrial laboratories, including: hypodermic needles, syringes, (with or without the attached needle), pasteur pipettes, scalpel blades, suture needles, blood vials, needles with attached tubing, and culture dishes (regardless of presence of infectious agents). Also included are other types of broken or unbroken glassware that were in contact with infectious agents, such as used slides and cover slips.

F.   Untreated biohazard waste should NEVER be disposed of in the municipal solid waste stream. All laboratories should evaluate their waste stream to ensure that all biohazard wastes, including sharps and syringes, are treated before disposal in the municipal waste stream. Prior to any treatment all biohazard wastes, including those to be incinerated, should be enclosed in a puncture-proof, BIOHAZARD BAG that is marked with the " biohazard waste'.

Chapter 4 - CHEMICAL SAFETY

Laboratory works involve handling of hazardous chemicals on a daily basis. To avoid any accident lab workers must have knowledge of general chemical safety, toxicological information, and procedures for handling and storage for the chemicals they are using. To keep employees safe this section addresses educational components and spells out specific protocols to minimize hazardous chemical exposures.

4.1 Modes of Entry

There are four major modes of entry to chemicals: inhalation, skin absorption, injection, and ingestion. Inhalation and skin absorption are the predominant occupational exposures you may expect to encounter in the laboratory and will be discussed in some detail. Accidental injection of chemicals can be eliminated by good laboratory safety practices. Accidental ingestion of chemicals can be eliminated by a combination of good laboratory and hygienic practices such as washing hands and prohibiting foods, drinks, cosmetics, and tobacco products in the laboratory workplace .All potential exposures are discussed in the Material Safety Data Sheets available for each chemical or product.

4.2 Basic Chemical Classifications

1. Volatile Solvents

Most ubiquitous in the laboratory setting. Generally subdivided into two categories: chlorinated (not flammable) and non-chlorinated (flammable). However, that the chlorinated solvents do decompose when burned generating toxic vapors.

If inhaled, produce drowsiness, dizziness and headaches cause upper respiratory and/or eye irritation and can lead to lung, liver ,and kidney damage.

Odor is not generally indicative of the degree of hazard that the material presents.

Skin absorption can cause drying and cracking of the skin, and may lead to chronic dermatitis with prolonged and repeated exposure.

Direct liquid contact by solvents in the eyes can be very serious. .

2. Acid and Bases

Have corrosive action on human tissues. Minor exposures are generally reversible yet painful.

3. Toxic Solids

In the form of a solution can penetrate the skin.

An oxidizing material dissolved in water can act directly on the skin causing irritation where the solid alone would be relatively less irritating.

In the solid form, the greatest risk of exposure is through inhalation.

4.3 Incompatible Chemicals

Certain hazardous chemicals cannot be mixed or stored safely with other chemicals due to potentially severe or extremely toxic reactions taking place. For example, keep oxidizing agents separated from reducing agents, initiators separated from monomers, and acids separated from alkalis, etc.

The chemical label and Material Safety Data Sheet contain information on incompatibilities.

A list of incompatible chemicals is included in Appendix B.

4.4 Chemical Stability

Stability refers to the susceptibility of the chemical to decomposition. Ethers, liquid paraffins, and olefins can form peroxides on exposure to air and light. Since these chemicals are packaged in an air atmosphere, peroxides can form even though the containers have remained sealed. Some inorganic chemicals also are unstable.

Examples of potential peroxide forming materials are included in Appendix A.

4.5 Shock-Sensitive Chemicals

Shock-sensitive refers to the sensitivity of the chemical to decompose rapidly or explode when struck, vibrated, or otherwise agitated.

The label and Material Safety Data Sheet will indicate if a chemical is shock-sensitive.

A partial list of potential shock-sensitive materials is included in Appendix C.

4.6 Material Safety Data Sheets

The Material Safety Data Sheet (MSDS) is a format for describing what chemical or product you are working with, potential chemical hazards, and ways of minimizing these hazards. These sheets should be on hand in the laboratory for people who use these chemicals. Information that is contained in the Material Safety Data Sheets is also required by law to be conveyed to employees on a chemical-by-chemical basis.

IT IS THE RESPONSIBILITY OF THE LABORATORY SUPERVISOR TO ENSURE THAT THE FACILITIES ARE ADEQUATE AND THAT THOSE WHO WILL HANDLE ANY MATERIAL HAVE RECEIVED PROPER TRAINING AND EDUCATION TO DO SO SAFELY.

4.7 Spill Prevention

For information on handling of chemical spills see - "Chemical Spills."

Proper techniques for transporting hazardous chemicals and proper storage techniques can help prevent spills. When large bottles of acids, solvents, or other liquids are transported within the laboratory without a cart, only one bottle should be carried at a time. The bottle should be carried with both hands, one on the neck of the bottle and the other underneath. Avoid the temptation to hook a finger through the glass ring on top of the bottle, allowing it to dangle while being transported. Never carry or attempt to pick up a bottle by the cap.

4.8 Chemical Storage

1. Every chemical in the laboratory should have a definite storage place and should be returned to that location after each use. Shelves should not be overloaded.

2. Utilize a compatible/suitable container for experiments, stored chemicals and collected wastes.

3. Caps and covers for containers should be securely placed whenever the container is not in immediate use.

4. Chemicals should be stored as close as feasible to the point of use in order to maximize efficiency and minimize transport distance. Small quantities of chemicals can be held at individual work stations if this quantity is to be promptly used in a test.

5. Out-of-date chemicals should be disposed of on a periodic basis to reduce overall hazard potential and minimize inventory tracking and updating.

4.9 CHEMICAL WASTE

Chemical waste is as harmful as chemicals in labs so these surpluses , expired or waste chemical material must be put in properly labeled containers at pre specified area. Label must have name of chemical and its hazards (flammable, toxic, corrosive etc).Containers must be in proper condition. Accumulation must not exceed available storage area.

CHAPTER 5 - RADIOLOGICAL SAFETY

Radioactive material are very harmful so special care should be taken if labs are using any radioactive material. Handling should be in such a way that it ensure least possible hazard to the individual worker, his associates, the University facilities, and the general public.

PERSONNEL MUST BE PROPERLY TRAINED FOR THE USE OF RADIOACTIVE MATERIALS.

5.1 Supervision of Usage and Disposal

- Usage and disposal of all radioisotopes and equipment should be in accordance with safety standards as are recommended by the National Bureau of Standards, the Oklahoma State Department of Health, and the Nuclear Regulatory Commission or its agencies, or as are established by the Committee.

- An area should be designated as a "Radiation Area" when a major portion of the body could receive any one hour dose in excess of 5 millirem, or in any five consecutive days a dose in excess of 100 millirem.

- An area should be designated as a "High Radiation Area" when a major portion of the body could receive in any one hour a dose in excess of 100 millirem.

5.2 Protective Clothing

Suitable gloves are recommended whenever hand contamination is probable. Rubber gloves should be worn when handling open vessels containing alpha material or when handling any equipment of comparable hazard. Rubber gloves are to be preferred for cases where liquid contamination may be present or where radioactive dust might filter through a cloth glove. Rubber gloves are to be cleaned, if practical, before removal. All gloves are to be stored and handled so as to prevent contamination of the inside surfaces.

Laboratory coats and/or aprons are recommended as additional protection for personal clothing during handling of materials where spillage is possible.

5.3 Contamination Control

a.   Accomplish decontamination before eating, smoking, applying makeup or leaving work.

b.   Wash rubber gloves before removing from hands unless the radiation level requires immediate removal.

c.   Refrain from wearing protective clothing outside of the laboratory area if there is any possibility it has been contaminated.

d.   If possible use available equipment to assure that decontamination has been effected.

e.   No work with long-lived alpha and beta-gamma emitters, in any chemical or physical form, is to be performed by a person having break in his skin below the wrist unless suitable gloves known to be clean on the inside are worn.

f.   The pipetting by mouth of liquids containing radioactivity is forbidden.

g.   All areas in which there is radiation that can pose threat should be physically isolated and appropriate signs posted to prevent persons from entering the area without being aware of the radiological hazard. Signs having the radiation symbol will be standard for radiation hazards.

h.   The symbol must also be used to distinguish radioactive source containers, contamination area, hot sinks, barriers, etc.

i.   All spills of radioactive material must be cleaned up promptly. Cleaning tools must not be removed or used elsewhere under any circumstances.

j.   Areas in which radioisotopes are used must be surveyed regularly using methods appropriate to the isotope(s) in use.

k.   Authorized users are responsible for the maintenance of up-to-date permanent records of the results of area surveys.

l.   Eating, smoking, or preparation of food and drink in a laboratory or room where radioactive materials are handled is not permitted.

m.   Smoking in such laboratories is also prohibited as is the application of cosmetics or any other activity which significantly increases inhalation and ingestion hazards.

n.   All packages must be examined carefully to ascertain if the package has been opened, damaged, and/or contaminated in transit.

o.   Due care should be exercised when opening packages of radioactive materials, including the wearing of gloves and other appropriate protective clothing and the use of hoods and other equipment to protect personnel and minimize the possibility of contamination.

p.   Waste disposal of radioactive material is itself an important part. Radioactive wastes produced must be stored on campus for radioactive decay or transported elsewhere for permanent disposal. Both the operations must be done in safe specified areas for solid, liquid or gaseous waste to avoid any sort of threat.

q.   It is essential to maintain record of waste disposal.

Chapter 6-Emergency

Your laboratory should have a plan for evacuation in case of an emergency. Do you know what your lab's emergency plan is for each of the following types of emergency?

- Fire
- Medical
- Chemical

Fire

You should only consider fighting a fire when all of the following statements are true:

- You have called the fire department and/or pulled the fire station lever.
- You have gotten everyone safely out of your laboratory and the building
- You have verified that the fire extinguisher available to you is full, in good condition, and is of the appropriate class to fight the fire.
- You have had training in the use of the fire extinguisher and are confident of your ability to use it properly.
- The fire is small and in a confined area such as a waste paper basket or hood.
- One or both of the fire exit doors will be located behind you when you face the fire in order to fight it with the extinguisher. If you have any doubts, exit the lab closing the door behind you and let the fire department, who are experts, do their job.

Fire exits

Where are the fire exits in your laboratory? There should be two clearly marked exits from each laboratory. These doors should not be blocked by furniture, equipment, or instrumentation.

Fire Extinguishers

Locate the fire extinguishers in your laboratory. What types of extinguishers do you have in your laboratory? Check to make sure that these extinguishers are the correct types for the kinds of hazards you are likely to face while working on your research project.

The fire extinguishers in your laboratory should be inspected on a regular basis by someone from either the Office of Environmental Health and Safety or Fire Safety at your institution. Don't make assumptions about safety equipment. Periodically check the date on the red tag and the gauge on the fire extinguisher to make sure that the extinguisher is full (gauge) and that the extinguisher is known to be in good working order. Always check these before using a fire extinguisher.

Types of Fire Extinguishers

There are four main types of fire extinguishers: A, B, C, and D.

- **Class A** fire extinguishers use water to put out paper and wood based fires.
- **Class B** fire extinguishers use compressed non-flammable gases such as carbon dioxide to put out fires involving flammable materials. The gas extinguishes the fire by starving it of oxygen. Note that these fire extinguishers should not be used in small confined spaces as they have the potential to asphyxiate the user, too, in the process.
- **Class C** fire extinguishers shoot a very fine non-flammable, non-conductive powder in order to extinguish electrical fires.
- **Class D** fire extinguishers are for use in combating fires involving flammable metals such as magnesium and sodium. These types of fires are especially dangerous. Unless you are trained, don't try to fight these fires.

There are also multi-class fire extinguishers as well. One of the most common multi-class fire extinguishers is the carbon dioxide extinguisher which can be used for Class B and C fires.

How to Properly Use a Fire Extinguisher

Fire extinguishers can be heavy and awkward to use effectively in an emergency situation if you aren't properly trained. If you haven't used a fire extinguisher before, it is really important to obtain training first.

PASS, which stands for pull, aim, squeeze, and sweep, is a common acronym used to summarize the general procedure for using a fire extinguisher properly:

- Pull the pin
- Aim the nozzle at the base of the fire
- Squeeze the handle and
- Sweep the spray across the base of the fire slowly back and forth until the fire is completely extinguished.

Don't walk away from the scene until you are certain that the fire has been completely extinguished.

Medical

First attempt is to ascertain the source of the problem. If the victim is unconscious, look around and make sure that electricity isn't responsible. If it is, use a non-conductive object to move the source of electricity away from the victim and seek immediate medical help.

If the victim is unconscious or does not appear to be breathing, request medical assistance immediately. Do not move the victim unless instructed to do so by medical personnel.

Chemical

If the victim appears to have been splashed with a chemical or solvent, assist them to the nearest emergency shower and pull the handle. Help the victim remove any contaminated clothing and be prepared to provide them with a clean lab coat or other temporary covering.

Emergency Contacts

Advance planning coupled with knowledge (information) is the best offense in case of an emergency. Locate the following information, insert it into the table provided, Xerox and paste the completed table publicly at your lab bench where you can see it in case of an emergency.

- Telephone number including area code
- Your workplace emergency phone number
- Your Advisor's home phone number
- Your personal physician

The equipment that should be available in your laboratory in case of emergency includes:

- Eye wash stations
- Showers
- Spill kits
- First aid kits
- Fire blankets

- Fire extinguishers
- Emergency exits

6.1 CHEMICAL SPILLS

In laboratories we are surrounded by different types of chemicals. Not necessarily we must be aware of properties of each of them but we should have knowledge about those that we use. **For safely using Chemical Reagents** always carefully read the reagent label and the material safety data sheet (MSDS).Issues to research and think carefully about before using a new reagent include the following:

- **Chemical compatibility** - Is this reagent known to be incompatible with any other reagents with which you or others in the laboratory might be working?
- **Chemical reactivity** - Is the reagent a strong oxidizer? Reductant? Does it react with moisture? Oxygen?
- **Flammability** - Is this reagent flammable?
- **Volatility** - Is this reagent volatile?
- **Toxicity** - Is the reagent toxic? Is it a mutagen? Carcinogen? What are the symptoms of exposure?
- **Handling** - What personal protective equipment should one use in working with this reagent? Gloves? What kind of gloves? Safety glasses? Should it be handled in a hood?
- **Accidents** - How should this material be cleaned up in case of a spill?
- **Emergencies** - What kinds of emergencies could arise from use/misuse of this chemical? Are you prepared to deal with these?

Chemicals have potential to harm handler, co-workers and laboratory and contaminate environment (water, air, soil).We should be prepare to take proper action in case of chemical spill. All chemicals are not hazardous so the steps need to be taken depends upon chemical's properties.

Spilled Liquids

Confine the liquid spill to a small area. For small quantities of inorganic acids or bases, use a neutralizing agent or an absorbent mixture (e.g., soda ash or diatomaceous earth). For small quantities of other materials, absorb the spill with a nonreactive material (such as vermiculite, clay, dry sand, or towels).For larger amounts of inorganic acids and bases, flush with large amounts of water (providing the water will not cause additional damage). Fooding is not recommended in storerooms where violent spattering may cause additional hazards or in areas where water-reactive chemicals may be present. Mop up the spill, wringing out the mop in a sink or a pail equipped with rollers. Carefully pick up and clean any cartons or bottles that have been splashed or immersed. If needed, vacuum the area with a HEPA filtered vacuum cleaner approved and designed for the material involved. If the spilled material is extremely volatile, let it evaporate and be exhausted by the laboratory hood (provided that the hood is authorized for use with the spilled chemical).

SPILLED SOLIDS

GENERALLY, SWEEP SPILLED SOLIDS OF LOW TOXICITY INTO A DUST PAN AND PLACE THEM INTO A CONTAINER SUITABLE FOR THAT CHEMICAL. ADDITIONAL PRECAUTIONS SUCH AS THE USE OF A VACUUM CLEANER EQUIPPED WITH A HEPA FILTER MAY BE NECESSARY WHEN CLEANING UP SPILLS OF MORE HIGHLY TOXIC SOLIDS. DISPOSE OF RESIDUES ACCORDING TO SAFE DISPOSAL PROCEDURES. REMEMBERING THAT PERSONAL PROTECTIVE EQUIPMENT, BROOMS, DUST PANS, AND OTHER ITEMS MAY REQUIRE SPECIAL DISPOSAL PROCEDURES. (SEE SECTION - "CHEMICAL WASTE"). REPORT THE CHEMICAL SPILL IN WRITING.

IT IS IMPORTANT TO BECOME INFORMED CONCERNING THE APPROPRIATE EMERGENCY PROTOCOLS FOR DEALING WITH WHATEVER ROUTINE HAZARDS YOU MAY ENCOUNTER WHILE WORKING IN YOUR LAB. DO YOU KNOW WHAT TO DO IN CASE OF AN EMERGENCY? IT IS CRITICAL TO LEARN WHAT THE APPROPRIATE EMERGENCY MEASURES ARE AND TO MAKE SURE YOU KNOW HOW TO USE THE AVAILABLE SAFETY EQUIPMENT NOW BECAUSE WHEN AN EMERGENCY ARISES THERE SIMPLY WON'T BE ANY TIME TO DO THIS. CHEMICAL SPILL CAN BE CLASSIFIED AS EMERGENCY SPILL IF IT-

- Causes personal injury or chemical exposure that requires medical attention;
- Causes a fire hazard or uncontrollable volatility;
- Requires a need for breathing apparatus of the supplied air or self-
- Contained type to handle the material involved;
- Involves or contaminates a public area;
- Causes airborne contamination that requires local or building evacuation;
- Causes a spill that cannot be controlled or isolated by laboratory personnel;
- Causes damage to property that will require repairs;
- Involves any quantity of metallic mercury;
- Cannot be properly handled due to lack of local trained personnel and/or equipment to perform a safe, effective cleanup;
- Requires prolonged or overnight cleanup;
- Involves an unknown substance; or
- Enters the land or water.

What should be done during an *emergency spill* ?

- ✓ Each spill incident is unique and involves persons with varying levels of spill expertise and experience. Don't get panic, make a call for help, isolate the spill and secure the area.

✓ If the spill presents an immediate danger, leave the spill site, warn others and evacuate them, control entry to the spill site, and wait for expertise help.

✓ Remove contaminated clothing. Flush skin/eyes with water at least 15 minutes to 30; use soap for intermediate and final cleaning of skin areas. Protect yourself (use personnel protective equipment).If required immediately seek medical attention.

✓ If flammable vapors are involved, do not operate electrical switches unless to turn off motorized equipment. Try to turn off or remove heat sources, where safe to do so.

✓ Do not touch the spill without protective clothing.

✓ If the substance involved is an unknown, then emergency spill response procedures are limited to self-protection.

✓ Where the spill does not present immediate personal danger, try to control the spread or volume of the spill. This could mean shutting a door, moving nearby equipment to prevent further contamination, repositioning an overturned container or one that has a hole in the bottom or side, creating a dike by putting an absorbent around a spill or opening the sashes on the fume hoods to facilitate removal of vapors.

✓ Never assume gases or vapors do not exist or are harmless because of lack of smell.

✓ Increase ventilation by opening closed fume hood sashes to the 12 inch or full open position. Exterior doors may be opened to ventilate non-toxic vapors.

✓ Use absorbents to collect substances. Reduce vapor concentrations by covering the surface of a liquid spill with absorbent. Control enlargement of the spill area by diking with absorbent.

6.2 Mercury Handling and Spill Clean Up

1. Health Effects

Metallic mercury can be absorbed into the body as well as through inhalation and ingestion into the skin. Mercury vapors are odorless, colorless, and tasteless. A quantity as small as 1 milliliter can evaporate over time, as raise levels in excess of allowable limits. Mercury poisoning from exposure by chronic inhalation can cause emotional disturbances, unsteadiness, inflammation of the mouth and gums, general fatigue, memory loss, and headaches. In most cases of exposure by chronic inhalation, the symptoms of poisoning gradually disappear when the source of exposure is removed. Improvement, however, may be slow and complete recovery may take years.

2. Storage and Handling

Because of the health effects of mercury, the extremely difficult and time-consuming procedures required to properly clean spills, every effort should be taken to prevent accidents involving mercury. Always store mercury in unbreakable containers and stored in a well-ventilated area. When breakage of instruments or apparatus containing mercury is a possibility, the equipment should be placed in an enameled or plastic tray or pan that can be cleaned easily and is large enough to contain the mercury. Transfers of mercury from one container to another should be carried out in a hood, over a tray or pan to confine any spills. If at all possible, the use of mercury thermometers should be avoided. If a mercury thermometer is required, many are now available with a Teflon® coating that will prevent shattering. Always wash hands after handling mercury to prevent skin absorption or irritation.

3. Air Monitoring

Any mercury spill has the potential to generate airborne concentrations in excess of regulated levels. Do air monitoring of the spill area before cleanup to determine the airborne concentration. Large spills or spills with elevated vapor levels may dictate cleanup by a qualified contractor.

4. Protective Clothing

For small spills, a laboratory coat, safety glasses, and gloves should be used. Gloves made of the following have been rated as excellent for protection against elemental mercury:

Chlorinated polyethylene (CPE),Polyvinyl Chloride (PVC), Polyurethane, Nitrile Rub ber(also known by Viton several brand names) ,Butyl Rubber,Neoprene

If mercury has been spilled on the floor, the workers involved in cleanup and decontamination should wear plastic shoe covers. EHS should be called immediately if a spill is extensive enough to require workers to kneel or sit where mercury has been spilled since Tyvek® or similar impermeable clothing will be required.

5. Spill Kits

Special spill kits are available from a variety of sources. If a spill kit is purchased, follow the manufacturer's directions. Alternatively, a kit can be assembled with the following components:

a.   protective gloves,

b.   mercury suction pump or disposable pipettes to recover small droplets,

c.   elemental zinc powder (or commercial amalgam material),

d.   dilute sulfuric acid (5-10%) in spray bottle,

e.   sponge or tool to work amalgam,

f.   plastic trash bag,

g.   plastic container (for amalgam), and

h.   plastic sealed vial for recovered mercury.

6. Clean Up Procedures

a. Wearing protective clothing, pools and droplets of metallic mercury can be pushed together and then collected by a suction pump.

b. After the gross contamination has been removed, sprinkler the entire area with zinc powder. Spray the zinc with the dilute sulfuric acid.

c. Using the sponge, work the zinc powder/sulfuric acid into a paste consistency while scrubbing the contaminated surface and cracks or crevices.

d. To minimize contamination of housekeeping items, stiff paper may be used to assist in cleaning up the amalgam.

e. After the paste has dried, it can be swept up and placed into the plastic container for disposal.

f. Rags, shoe covers, sponges, and anything used for the cleanup should be placed in the trash bag to be disposed of as contaminated material.

7. Waste Disposal

Appropriate removal of the mercury waste and contaminated items is necessary.

6.3 RADIATION SPILLS

More hazardous material you are dealing with more careful you should be. In the event of dissemination of radioactive materials, the following general procedures are given as a guide to be adapted to the specific nature of the emergency.

A.   Minor Spills Involving No Radiation Hazard to Personnel

✓ Notify all other persons in the room at once.

✓ Permit only the minimum number of persons necessary to deal with the   spill into the area.

✓ Confine the spill immediately.

Liquid Spills:

a.   *Don protective gloves.*

b.   *Drop absorbent paper on the spill.*

Dry Spills:

a.   *Don protective gloves.*

b.   *Dampen thoroughly, taking care not to spread the contamination.*

✓ Notify the faculty member in charge of the laboratory and the Radiological Safety Officer (if appointed) as soon as possible.
✓ Monitor all persons involved in the spill and cleaning.

✓ Decontaminate the area according to the directions of the Campus Radiological Safety Officer.

✓ Permit no person to resume work in the area until a survey is made, and approval of the Radiological Safety Officer is secured.

✓ Prepare a complete history of the accident and subsequent activity related thereto for the records of the Radiological Safety Officer.

B.   Major Spills Involving Radiation Hazard to Personnel

- ✓ Notify all persons not involved in the spill to vacate the room at once.

- ✓ If the spill is a liquid, and the hands are protected, right the container.

- ✓ If the spill is on the skin, flush thoroughly.

- ✓ If the spill is on clothing, discard outer or protective clothing at once.

- ✓ Shut off air conditioning units serving the laboratory.

- ✓ Vacate the room.

- ✓ Notify the faculty member in charge and the Radiological Safety Officer as soon as possible.

- ✓ Take immediate steps to decontaminate personnel involved, as necessary.

- ✓ Decontaminate the area per the recommendations of the Radiological Safety Officer. (Personnel involved in decontamination must be adequately protected.)

- ✓ Monitor all persons involved in the spill and cleaning to determine adequacy of decontamination.

- ✓ Permit no person to resume work in the area until a survey is made and approval of the Radiological Safety Officer is secured.

- ✓ Prepare a complete history of the accident and subsequent activity related thereto for the records of the Radiological Safety Officer.

C. **Accidents Involving Radioactive Dusts, Mists, Fumes, Organic Vapors, and Gases**

- ✓ Notify all other persons to vacate the room immediately.

- ✓ Hold breath and vacate room.

- ✓ Shut off air conditioning by master switch.

- ✓ Notify the faculty member in charge and the radiological safety officer at once.

✓ Ascertain that all doors giving access to the room are closed and post conspicuous warnings or guards to prevent accidental opening of doors.

✓ Report at once all known or suspected inhalations of radioactive materials.

✓ The Radiological Safety Officer shall evaluate the hazard and the necessary safety devices for safe re-entry.

✓ Determine the cause of contamination and rectify the condition.

✓ Decontaminate the area.

✓ Perform air survey of the area before permitting work to be resumed.

✓ Monitor all persons suspected of contamination.

✓ Prepare a complete history of the accident and subsequent activity related thereto for the records of the Radiological Safety Officer.

D. **Injuries to Personnel Involving Radiation Hazard**

✓ Wash minor wounds immediately, under running water, while spreading the edges of the gash. Consult physician.

✓ Report all radiation accidents to personnel (wounds, overexposure, ingestion, inhalation) to the faculty member in charge and the Radiological Safety Officer as soon as possible.

✓ Permit no person involved in a radiation injury to return to work without the approval of the Radiological Safety Officer and attendant physician.

✓ Prepare a complete history of the accident and subsequent activity related thereto for the records of the Radiological Safety Officer.

6.4 BIOHAZARD SPILLS

A. Biological Spills

Biological spills outside biological safety cabinets generate aerosols that can be dispersed in the air throughout the laboratory. These spills can be very serious if they involve microorganisms that require Biosafety Level 3 containment, since most of these agents have the potential for transmitting disease by infectious aerosols. To reduce the risk of inhalation exposure in such an accident, occupants should leave the laboratory immediately. The laboratory should not be reentered to decontaminate or clean up the spill for at least 30 minutes. During this time the aerosol may be removed from the laboratory via the exhaust ventilation systems, such as biological safety cabinets or chemical fume hoods, if present.

1. Spills on the Body

   ✓ Remove contaminated clothing.
   ✓ Vigorously wash exposed area with soap and water for one minute.
   ✓ Obtain medical attention (if necessary).
   ✓ Report the incident to the laboratory supervisor.

2. Biosafety Level 1 Organism Spill

   ✓ Wear disposable gloves.
   ✓ Soak paper towels in disinfectant and place over sill.
   ✓ Place towels in a plastic bag for disposal.
   ✓ Clean up spill area with fresh towels soaked in disinfectant.

3. Biosafety Level 2 Organism Spill

   ✓ Alert people in immediate area of spill.
   ✓ Put on protective equipment. This may include a laboratory coat with long sleeves, back-fastening gown or jumpsuit, disposable gloves, disposable shoe covers, safety goggles, mask or full-face shield.
   ✓ Cover spill with paper towels or other absorbent materials.
   ✓ Carefully pour a freshly prepared 1 to 10 dilution of household bleach around the edges of the spill and then into the spill. Avoid splashing.
   ✓ Allow a 20-minute contact period.
   ✓ After the spill has been absorbed, clean up the spill area with fresh towels soaked in disinfectant.

✓ Place towels in a plastic bag and decontaminate in an autoclave.

4. Biosafety Level 3 Organism Spill

✓ Attend to injured or contaminated persons and remove them from exposure.
✓ Alert people in the laboratory to evacuate.
✓ Close doors to affected area.
✓ Call for campus emergency response.
✓ Have a person knowledgeable of the incident and laboratory assist emergency personnel when they arrive.

B. Blood Spills

Cleaning of blood spills should be limited to those persons who are trained for the task. If an untrained person encounters a spill, he/she should limit access to the area and immediately call the person(s) assigned to this duty. Only disposable towels should be used to avoid the difficulties involved in laundering. If a spill involves broken glassware, the glass should never be picked up directly with the hands. It must be cleaned up using mechanical means, such as a brush and dustpan, tongs, or forceps. Furthermore,

✓ Persons who clean blood spills should wear disposable gloves of sufficient strength so they will not tear during cleaning activities. If the gloves develop holes, tears, or splits, remove them, wash hands immediately, and put on fresh gloves. Disposable gloves must never be washed or reused. Remove gloves one at a time by grasping the wrist opening and pulling toward the fingers so that the gloves come off as inside out. Double-bag gloves with other contaminated biomedical waste (such as towels).

✓ If enough blood has been spilled to expect splashing during cleaning, call EHS. Additional protective equipment may be required. EHS can provide a face-shield and other protective clothing that your staff can use if splashing is expected.

Disinfectants

Read and follow all manufacturer's handling instructions. All spills of blood and blood-contaminated fluids should be properly cleaned using any of disinfectants. A solution of 5.25 percent sodium hypochlorite (household bleach) diluted between 1:10 and 1:100 with water (a 1:100 dilution of common household bleach yields 500 parts per million free available chlorine - approximately ¼ cup of bleach per gallon of tap water).

Cleaning Blood Spills on Hard Surfaces

To assure the effectiveness of any sterilization or disinfection process, surfaces must first be thoroughly cleaned of all visible blood or soil before a germicidal chemical is applied for disinfection.

- ✓ Isolate the area, if possible.

- ✓ Wear gloves and other protective apparel as needed.

- ✓ Remove visible blood with disposable towels in a manner that will ensure against direct contact with the blood. For example, put towels over the spill to absorb the liquid.

- ✓ Place contaminated towels in a plastic waste disposal bag.

- ✓ The area should then be decontaminated with an appropriate germicide applied according to manufacturer's directions.

- ✓ All contaminated towels and gloves should be double-bagged for disposal and labeled with the biohazard symbol.

Cleaning Blood Spills on Carpeting

- ✓ Use only a registered germicide. Read and follow manufacturer's instructions. Do not use chlorine bleach solution on carpet.

- ✓ Isolate the area--if possible.

- ✓ Wear gloves and other appropriate apparel.

- ✓ Procedures for small spills on carpets (smaller than a quarter) are as follows.

  - Soak the spill with enough disinfectant to cover the spot.

  - Let dry at least overnight to ensure that the spot is disinfected.

  - Shampoo carpet, if needed, or use 3% hydrogen peroxide to remove discoloration.

- ✓ Procedures for larger spills are as follows.

  - Pour disinfectant on the spot and let stand at least 30 minutes to allow some disinfection to take place. Blot up excess liquid with disposable towels.

  - Soak the area with additional disinfectant. Allow to dry overnight. Shampoo carpet, if needed, or use 3% hydrogen peroxide to remove discoloration.

✓ All contaminated towels and gloves should be double-bagged and labeled with the biohazard symbol.

C. Cytotoxic/Antineoplastic Spills

1. General Procedures

    ✓ Spills and breakages of cytotoxic/antineoplastic drugs (CDs) should be cleaned up immediately by a properly trained person using the appropriate procedures.
    ✓ Broken glass should be carefully removed.
    ✓ A spill should be identified with a warning sign so that other persons in the area will not be contaminated.

2. Personnel Contamination

    ✓ Overt contamination of gloves or gowns, or direct skin or eye contact should be treated as follows.
    ✓ Immediately remove the gloves or gown.
    ✓ Wash the affected skin area immediately with soap (not germicidal cleanser) and water. For eye exposure, immediately flood the affected eye with water or isotonic eyewash designated for the purpose for at least five minutes.
    ✓ Obtain medical attention immediately.

3. Clean-up of Small Spills

    Spills of less than 5 ml or 5 gm outside a hood should be cleaned immediately by personnel wearing gowns, double surgical latex gloves, and eye protection.

    ✓ Liquids should be wiped with absorbent gauze pads, solids should be wiped with wet absorbent gauze. The spill areas then should be cleaned (three times) using a detergent solution followed by clean water.
    ✓ Any broken glass fragments should be placed in a small cardboard or plastic container and then into a CD disposal bag, along with the used absorbent pads and any non-cleanable contaminated items.
    ✓ Reusable glassware or other contaminated items should be placed in a plastic bag and washed in a sink with detergent by a trained employee wearing double surgical latex gloves.

4. Clean-up of Large Spills

For spills of amounts larger than 5 ml or 5 gm, the spread should be limited by gently covering with absorbent sheets of spill-control pads or pillows or, if a powder is involved, with damp cloths or towels. Be sure not to generate aerosols. Access to the spill areas should be restricted.

- ✓ Protective apparel should be used with the addition of a respirator when there is any danger of airborne powder or an aerosol being generated. The dispersal of CD particles into surrounding air and the possibility of inhalation is a serious matter and should be treated as such.
- ✓ Chemical inactivators, with the exception of sodium thiosulfate, which can be used safely to inactivate nitrogen mustard, may produce hazardous by-products and should not be applied to the spilled drug.
- ✓ All contaminated surfaces should be thoroughly cleaned with detergent solution and then wiped with clean water. All contaminated absorbents and other materials should be disposed of in the CD disposal bag.

5.   Spills in Hoods

If the spill occurred in either a glove box, clean bench or biological safety cabinet, the HEPA filter (contained in the cabinet) is more than likely contaminated. Label the unit "Do Not Use--Contaminated With (name of substance)." The HEPA filter and filter cabinet must be decontaminated and the filter changed and properly disposed of. This procedure may require the services of an outside contractor trained in the use of specialized personal protective equipment.

6.   Spill Kits

Spill kits, clearly labeled, should be kept in or near preparation and administrative areas. It is suggested that kits include a respirator, chemical splash goggles, two pairs of gloves, two sheets (12x12) of absorbent material, 250 ml and one liter spill control pillows and a small scoop to collect glass fragments. Absorbents should be suitable for incineration. Finally, the kit should contain two large CD waste-disposal bags.

7.   Waste Disposal

Disposal of all Communicable Diseases contaminated materials must be arranged through biohazard disposal unit of Institute.

SECTION 1.4 - LEAKING COMPRESSED GAS CYLINDERS

If a leak is suspected in cylinder such as in valve threads, valve stem and safety device , do not use a flame for detection; rather, a flammable-gas leak detector or soapy water or other suitable "snoop" solution should be used. If the leak cannot be remedied by tightening a valve gland or a packing nut, emergency action procedures should be affected. Laboratory workers should never attempt to repair a leak at the valve threads or safety device; rather, they should consult with the supplier for instructions.

If the substance in the compressed gas cylinder is not inert, or is hazardous, then use the procedures in Section - "Chemical Spills".

If the substance in the compressed gas cylinder is inert, or non-hazardous, contact the supplier for instructions.

SECTION 1.5 - FIRES

Fires are a common emergency in a chemistry laboratory.

In the event of a fire, do the following things:

A.   Assist any person in immediate danger to safety, if it can be accomplished without risk to yourself.

B.   Immediately activate the building fire alarm system.

C.   If the fire is small enough, use a nearby fire extinguisher to control and extinguish the fire. Don't fight the fire if these conditions exist:

      a. The fire is too large or out of control.

      b. If the atmosphere is toxic.

D.   If the first attempts to put out the fire do not succeed, evacuate the building immediately. Do not use elevators; use building stairwells

E.   Doors, and if possible, windows, should be closed as the last person leaves a room or area of a lab.   .

F.   Upon evacuating the building, personnel shall proceed to the designated meeting area (at least 150 feet from the affected building) where the supervisors are responsible for taking a head count and accounting for all personnel. No personnel to be allowed to re-enter the building without permission.

J.   All fires must be reported and investigated.

SECTION 1.6 - MEDICAL EMERGENCIES

Personal injuries are usually minor cuts or burns but can be as severe as acute effects of chemical exposure or incidents such as heart attacks or strokes.

The initial responsibility for first aid rests with the first person(s) at the scene, who should react quickly but in a calm and reassuring manner. The person assuming responsibility should immediately summon medical help (be explicit in reporting suspected types of injury or illness, location of victim, and type of assistance required). Send people to meet the ambulance crew at likely entrances of the building. The injured person should not be moved except where necessary to prevent further injury.

- The number to call for medical emergencies should be posted by your telephone.
- Detail of all such incidents must be reported down as mentioned.

Prevention of injuries should be a major emphasis of any laboratory safety program. Proper training will help prevent injuries from glassware, toxic chemicals, burns and electrical shock. In the event of any type of injury beyond that which first aid can treat, call for medical assistance. A minor injury may indicate a hazardous situation which should be corrected to prevent a serious future injury. It is important to document a minor injury as having been "work related" if the injury later leads to serious complications, such as from an infected cut.

Personal Protection During First Aid

When employees respond to emergencies they should take precautions to avoid exposure to blood and other potentially infectious materials such as gloves, masks, and protective clothing, which provide a barrier between the worker and the exposure source. For most situations in which first aid is given, the following guidelines should be adequate.

a.   For bleeding control with minimal bleeding and for handling and cleaning instruments with microbial contamination, disposable gloves alone should be sufficient.

57

b.	For bleeding control with spurting blood, disposable gloves, a gown, a mask, and protective eye wear are recommended.

c.	For measuring temperature or measuring blood pressure, no protection is required.

After emergency care has been administered, hands and other skin surfaces should be washed immediately and thoroughly with warm water and soap if contaminated with blood, other body fluids to which universal precautions apply, or potentially contaminated articles. Hands should always be washed after gloves are removed, even if the gloves appear to be intact.

SECTION 1.7 - ACCIDENT REPORTING

Employees should understand that the purpose of reporting and documenting accidents is not to affix blame, but instead to determine the cause of the accident so that similar incidents may be prevented in the future.

APPENDIX A

POTENTIAL PEROXIDE-FORMING CHEMICALS[1]

Acetal

Cyclohexene

Decahydronaphthalene

Diacetylene

Dicyclopentadiene

Diethyl Ether

Diethylene Glycol

Dimethyl Ether

para-Dioxane

Divinyl Acetylene

Ether (Glyme)

Ethylene Glycol Dimethyl Ether

Tetrahydronaphthalene

Methyl Acetylene

Isopropyl Ether

Tetrahydrofuran

Sodium Amide

Vinyl Ethers

Vinylidene Chloride

---

[1]   From Manufacturing Chemists' Association, Guide for Safety in the Chemical Laboratory, pages 215-217.

APPENDIX B

INCOMPATIBLE CHEMICALS[2]

Chemical	Keep out of Contact With:
Acetic Acid	Nitric acid, hydroxyl compounds, ethylene glycol, perchloric acid, peroxides, permanganates
Acetylene	Chlorine, bromine, copper, fluorine, silver, mercury
Alkali Metals	Water, carbon tetrachloride or other chlorinated hydrocarbons, carbon dioxide, the halogens
Ammonia, Anhydrous	Mercury, chlorine, calcium hypochlorite, iodine, bromine, hydrofluoric acid
Ammonium nitrate	Acids, metal powders, flammable liquids, chlorates, nitrites, sulfur, finely divided organic or combustible materials
Aniline	Nitric acid, hydrogen peroxide
Bromine	Same as chlorine: ammonia, acetylene, butadiene, butane, methane, propane (or other petroleum gases), hydrogen, sodium carbide, turpentine, benzene, finely divided metals
Butyl lithium	Water.
Carbon, activated	Calcium hypochlorite, all oxidizing agents
Chlorates	Ammonium salts, acids, metal powders, sulfur, finely divided organic or combustible materials
Chromic Acid	Naphthalene, camphor, glycerin, turpentine, alcohol, flammable liquids in general
Chlorine	Same as bromine: ammonia, acetylene, butadiene, butane, methane, propane (or other petroleum gases), hydrogen, sodium carbide, turpentine, benzene, finely divided metals
Chlorine dioxide	Ammonia, methane, phosphine, hydrogen sulfide
Copper	Acetylene, hydrogen peroxide
Cumene hydroperoxide	Acids, organic or inorganic
Cyanides (Na, K)	Acids
Flammable liquids	Ammonium nitrate, chromic acid, hydrogen peroxide, nitric acid, sodium peroxide, halogens, other oxidizing agents

---

[2]   From Manufacturing Chemists' Association, <u>Guide for Safety in the Chemical Laboratory</u>, pages 215-217.

Hydrocarbons	Fluorine, chlorine, bromine, chromic acid, sodium peroxide
Hydrocyanic acid	Nitric acid, alkalis
Hydrofluoric acid	Ammonia, aqueous or anhydrous
Hydrogen peroxide	Copper, chromium, iron, most metals or their salts, alcohols, acetone, organic materials, aniline, nitromethane, flammable liquids, oxidizing gases
Hydrogen sulfide	Fuming nitric acid, oxidizing gases
Iodine	Acetylene, ammonia (aqueous or anhydrous), hydrogen
Mercury	Acetylene, fulminic acid, ammonia
Nitric Acid	Acetic acid, aniline, chromic acid, hydrocyanic acid, hydrogen sulfide, flammable liquids, flammable gases
Oxalic acid	Silver, mercury
Perchloric acid	Acetic anhydride, bismuth and its alloys, alcohol, paper, wood, sulfuric acid, organics
Potassium	Carbon tetrachloride, carbon dioxide, water
Potassium permanganate	Glycerin, ethylene glycol, benzaldehyde, sulfuric acid
Silver	Acetylene, oxalic acid, tartaric acid, ammonium compounds
Sodium	Carbon tetrachloride, carbon dioxide, water
Sodium peroxide	Ethyl or methyl alcohol, glacial acetic acid, acetic anhydride, benzaldehyde, carbon disulfide, glycerin, ethylene glycol, ethyl acetate, methyl acetate, furfural
Sulfuric acid	Potassium chlorate, potassium perchlorate, potassium permanganate (or compounds with similar light metals, such as sodium, lithium, etc.)

APPENDIX C

<u>POTENTIAL SHOCK-SENSITIVE CHEMICALS</u>[3]

Acetylides of heavy metals	Fulminate of silver
Aluminum ophorite explosive	Fulminating gold
Amatol explosive (sodium amatol)	Fulminating mercury
Ammonal	Fulminating silver
Ammonium nitrate	Fulminating platinum
Ammonium perchlorate	Gelatinized nitrocellulose
Ammonium picrate	Guanyl nitrosamino guanyl tetrazene
Ammonium salt lattice	Guanyl nitrosamino guanylide hydrazine
Calcium nitrate	Heavy metal azides
Copper Acetylide	Hexanite
Cyanuric triazide	Hexanitrodiphenylamine
Cyclotrimethylenetrinitramine	Hexanitrostilbene
Cyclotetramethylenetranitramine	Hexogen (Cylclotrimethylenetrinitramine)
Dinitroethyleneurea	Hydrazoic acid
Dinitroglycerine	Lead azide
Dinitrophenol	Lead mannite
Dinitrophenolates	Lead picrate
Dinitrophenyl hydrazine	Lead salts
Dinitroresorcinol	Lead styphnate
Dinitrotoluene	Magnesium ophorite
Dipicryl sulfone	Mannitol hexanitrate
Dipicrylamine	Mercury oxalate
Erythritol tetranitrate	Mercury tartrate
Fulminate of mercury	Mononitrotoluene
Nitrated carbohydrate	Silver styphnate
Nitrated glucoside	Silver tetrazene

---

[3]   From Manufacturing Chemists' Association, <u>Guide for Safety in the Chemical Laboratory</u>, pages 215-217.

Nitrated polyhydric alcohol

Nitrogen trichloride

Nitrogen triiodide

Nitroglycerin

Nitroglycol

Nitroguanidine

Nitroparaffins

Nitromethane

Nitronium perchlorate

Nitrourea

Organic amine nitrates

Organic nitramines

Organic peroxides

Picramic acid

Picramide

Picratol explosive (ammonium picrate)

Picric acid

Picryl chloride

Picryl fluoride

Polynitro aliphatic compounds

Potassium nitroaminotetrazole

Silver acetylide

Silver azide

Sodatol

Sodium amatol

Sodium dinitro-ortho-cresolate

Sodium nitrate-potassium nitrate explosive mixtures

Sodium picramate

Styphnic acid

Tetrazene (guanyl nitrosamino guanyl tetrazene)

Tetranitrocarbazole

Tetrytol

Trimethylolethane

Trimonite

Trinitroanisole

Trinitrobenzene

Trinitrobenzoic acid

Trinitrocresol

Trinitro-meta-cresol

Trinitronaphthalene

Trinitrophenol

Trinitrophloroglucinol

Trinitroresorcinol

Tritronal

Urea nitrate

BIBLIOGRAPHY

Portions of this Manual were adapted with permission from
Young, J.A.; Kingsley, W.K.; Wahl, Jr., G.H.
Developing a Chemical Hygiene Plan;
American Chemical Society, Washington, DC, 1990; pp10-30.
Copyright 1990 American Chemical Society.

Informing Workers of Chemical Hazards,
American Chemical Society, Washington, DC (1985)

Chemical Safety Manual for Small Business,
American Chemical Society, Washington, DC (1989)

1990 Emergency Response Guidebook,
U.S. Department of Transportation, (1990)

Prudent Practices for Handling Hazardous Chemicals
National Academy Press, Washington, DC (1980)

Additional Sources of Information

Armour, M.A.,L.M. Browne, and GIL. Weir.
Hazardous Chemicals: Information and Disposal Guide,
3rd ed. Edmonton, Alberta: University of Alberta, 1987. 463 pp.

Armour, M.A., et al. *Potentially Carcinogenic Chemicals: Information and Disposal Guide*.
Edmonton, Alberta: University of Alberta, 1990(?). 147 pp.

Flinn Scientific, *Chemical Catalog/Reference Manual*, annual.

Safety in the Chemical Laboratory, vols 2, 3, and 4. *Journal of Chemical Education*.

Sax, N. Irving and Richard J. Lewis, Sr. *Hazardous Chemicals Desk Reference*, 2nd ed.
New York: Van Nostrand Reinhold, 1990. 1100 pp.

The Merck Index: An Encyclopedia of Chemicals, Drugs, and Biologicals.
Rahwey, NJ: Merck and Co., 1989.

National Fire Protection Association, *National Fire Codes*
11 vols with index. 1991

Sax, Irving N., *Dangerous Properties of Industrial Materials*,
6th ed. New York: Van Nostrand Reinhold Company, 1984 (3164 pp).

Weiss, G., ed. *Hazardous Chemicals Data Book*. Park Ridge, NJ: Noyes Data Corp., 1986

GLOSSARY

AA	Atomic absorption spectrophotometers
Action level	A concentration for a specific substance, calculated as an eight (8) hour time-weighted average, which initiates certain required activities such as exposure monitoring and medical surveillance. Typically it is one-half that of the PEL for that substance.
ACGIH	The American Conference of Governmental Industrial Hygienists is a voluntary organization of professional industrial hygiene personnel in government or educational institutions. The ACGIH develops and publishes recommended occupational exposure limits each year called Threshold Limit Values (TLVs) for hundreds of chemicals, physical agents, and biological exposure indices.
Acute Exposure	Single exposure episodes which occur over a short time period.
ANSI	The American National Standards Institute is a voluntary membership organization (run with private funding) that develops consensus standards nationally for a wide variety of devices and procedures.
ASHRAE	American Society of Heating, Refrigeration and Air Conditioning Engineers
Asphyxiant	A chemical (gas or vapor) that can cause death or unconsciousness by suffocation. Simple asphyxiants such as nitrogen, either use up or displace oxygen in the air. They become especially dangerous in confined or enclosed spaces. Chemical asphyxiants, such as carbon monoxide and hydrogen sulfide, interfere with the body's ability to absorb or transport oxygen to the tissues.
Biohazard Wastes	Discarded materials "that are biological agents or conditions (as an infectious organism or unsecure laboratory condition) that constitutes a hazard to man or his environment." This definition includes "any and all substances which contain materials to which organisms may cause injury or disease to man or his environment, but which are not regulated as controlled industrial waste."
BSC	Biological Safety Committee
BSO	Biological Safety Officer
"C" or Ceiling	A description usually seen in connection with a published exposure limit. It refers to the concentration that should not be exceeded, even for an instant. It may be written as TLV-C or Threshold Limit Value--Ceiling. (See also Threshold Limit Value).
Carcinogen	Any substance that causes the development of cancerous growths in living tissue, either those that are known to induce cancer in man or animals or experimental carcinogens that have been found to cause cancer in animals under experimental conditions.
C.A.S. Number	Identifies a particular chemical by the Chemical Abstracts Service, a service of the American Chemical Society that indexes and compiles abstracts of worldwide chemical literature called "Chemical Abstracts". These numbers are always contained in brackets.
CDC	Centers for Disease Control
CFR	Code of Federal Regulations

Chemical Hygiene Plan	A written program developed and implemented by the employer which sets forth procedures, equipment, personal protective equipment and work practices that are capable of protecting employees from the health hazards presented by hazardous chemicals used in that particular work place and meets the requirements of 29 CFR 1910.1450(e).
CHP	See Chemical Hygiene Plan
Chemical Reaction	A change in the arrangement of atoms or molecules to yield substances of different composition and properties. (See Reactivity).
Chronic Exposure	A series of exposures occurring over a longer period of time.
Combustible	A combustible liquid or an "Ordinary Combustible" such as wood, paper, etc.
Combustible Liquid	Any liquid having a flashpoint at or above 100 ^{O}F (37.8 ^{O}C), but below 200 ^{O}F (93.3 ^{O}C), except any mixture having components with flashpoints of 200 ^{O}F (93.3 ^{O}C), or higher, the total volume of which make up 99 percent or more of the total volume of the mixture.
Corrosive	Any gas, liquid, or solid that causes destruction of human tissue or a liquid that has a severe corrosion rate on steel. Generally, a substance that has a very low or a very high pH.
Cutaneous	Pertain to or affecting the skin.
Decomposition	The breakdown of a chemical or substance into different parts or simpler compounds. Decomposition can occur due to heat, chemical reaction, decay, etc.
Dermal	Pertaining to or affecting the skin.
Dermatitis	An inflammation of the skin.
Designated Area	An area which may be used for work with "select carcinogens, reproductive toxins, or substances which have a high degree of acute toxicity." A designated area may be the entire laboratory, an area of a laboratory, or a device such as a laboratory hood. A designated area shall be placarded to reflect the designated hazard.
Dose	The concentration of a substance and the time period during which the exposure occurs. The dose received links hazard and toxicity.
DOT	The United States Department of Transportation is the federal agency that regulates the labeling and transportation of hazardous materials.
Dyspnea	Shortness of breath; difficult or labored breathing.
Emergency Spills	Accidental chemical discharges that present an immediate danger to personnel and/or the environment. Under these circumstances, leave the spill site immediately and send for help. Management of these spills is the responsibility of specially trained and equipped personnel. Contact the campus police at 911 for response. They will notify the appropriate persons/departments. (See Section 1.1 - "Chemical Spills")
Employee	An individual employed in a laboratory work place who may be exposed to hazardous materials in the course of his or her assignments.

EPA	The Environmental Protection Agency is the governmental agency responsible for administration of laws to control and/or reduce pollution of air, water, and land systems.
EPA Number	The number assigned to chemicals regulated by the Environmental Protection Agency (EPA).
Erythema	A reddening of the skin.
Fires	
Class A	Fires in ordinary combustible materials such as wood, cloth, paper, rubber, and many plastics.
Class B	Fires in flammable liquids, oils, greases, tars, oil-base paints, lacquers and flammable gases.
Class C	Fires that involve energized electrical equipment where the electrical conductivity of the extinguishing medium is of importance; when electrical equipment is de-energized, extinguishers for class A or B fires may be safely used.
Class D	Fires in combustible metals such as potassium, sodium, lithium, magnesium, titanium, sirconium.
Flammable	Any substance which may be classified as a flammable aerosol, flammable gas, flammable liquid or flammable solid.
Flammable Aerosol	An aerosol that, when tested by the method described in 16 CFR 1500.45, yields a flame protection exceeding 18 inches at full valve opening, or a flashback (a flame extending back to the valve) at any degree of valve opening.
Flammable Gas	A gas that, at ambient temperature and pressure, forms a flammable mixture with air at a concentration of 13 percent by volume or less; or a gas that, at ambient temperature and pressure, forms a range of flammable mixtures with air wider that 12 percent by volume, regardless of the lower limit.
Flammable Liquid	Any liquid having a flashpoint below 100°F (37.8°C), except any mixture having components with flashpoints of 100°F (37.8°C), or higher, the total volume of which make up 99 percent or more of the total volume of the mixture.
Flammable Solid	A solid, other than a blasting agent or explosive, that is liable to cause fires through friction, absorption of moisture, spontaneous chemical change, retained heat from processing, or which can be ignited readily, and when ignited burns so vigorously and persistently as to create a serious hazard. A chemical shall be considered a flammable solid if, when tested by the method described in 16 CFR 1500.44, it ignites and burns with a self-sustained flame at a rate greater than one-tenth of an inch per second along its major axis.
Hazard	The possibility that exposure to a substance will cause injury when a specific quantity is used under certain conditions.

Health Hazard	A substance for which there is statistically significant evidence based on at least one study conducted in accordance with established scientific principles that acute or chronic health effects may occur in exposed employees. This term includes carcinogens, toxic or highly toxic agents, reproductive toxins, irritants, corrosives, sensitizers, hepatotoxins, nephrotoxins, neurotoxins, agents which act on the hematopoietic systems, and agents which damage the lungs, skin, eyes, or mucous membranes.
ICP	Inductively-coupled argon spectrometers
IDLH	Immediately dangerous to life or health concentrations represent the maximum concentration from which one could escape within 30 minutes without a respirator and without experiencing any escape-impairing (e.g., severe eye irritation) or irreversible health effects.
Ignitable	A solid, liquid, or compressed gas that has a flashpoint of less $140^{\circ}F$. Ignitable material may be regulated by the EPA as a hazardous waste, as well.
Incompatible	The term applied to two substances to indicate that one material cannot be mixed with the other without the possibility of a dangerous reaction.
Ingestion	Taking a substance into the body through the mouth as food, drink, medicine, or unknowingly as on contaminated hands or cigarettes, etc.
Inhalation	The breathing in of an airborne substance that may be in the form of gases, fumes, mists, vapors, dusts, or aerosols.
Inhibitor	A substance that is added to another to prevent or slow down an unwanted reaction or change.
Irritant	A substance that produces an irritating effect when it contacts skin, eyes, nose, or respiratory system.
LC_{50}	See Lethal Concentration$_{50}$.
LD_{50}	See Lethal Dose$_{50}$.
LEL	See Lower Explosive Limit.
Lethal Concentration$_{50}$	The concentration of an air contaminant that will kill 50 percent of the test animals in a group during a single exposure.
Lethal Dose$_{50}$	The dose of a substance or chemical that will kill 50 percent of the test animals in a group within the first 30 days following exposure.
LFL	See Lower Explosive Limit.
Lower Explosive Limit	(Also known as Lower Flammable Limit.) The lowest concentration of a substance that will produce a fire or flash when an ignition source (flame, spark, etc.) is present. It is expressed in percent of vapor or gas in the air by volume. Below the LEL or LFL, the air/contaminant mixture is theoretically too "lean" to burn. (See also UEL.)
Minor Spills	Small chemical leaks that usually are detected early and present no immediate danger to personnel or the environment. These are spills that can be safely corrected with the advice of knowledgeable laboratory or supervisory personnel.

MSDS	Material Safety Data Sheet (See Section 9.3)
Mutagen	Anything that can cause a change (or mutation) in the genetic material of a living cell.
Narcosis	Stupor or unconsciousness caused by exposure to a chemical.
NFPA	The National Fire Protection Association is a voluntary membership organization whose aims are to promote and improve fire protection and prevention. NFPA has published several volumes of codes known as the National Fire Codes.
NIH	National Institute of Health
NIOSH	The National Institute for Occupational Safety and Health is a federal agency that among its various responsibilities trains occupational health and safety professionals, conducts research on health and safety concerns, and test and certifies respirators for work place use.
Odor Threshold	The minimum concentration of a substance in the air at which a majority of test subjects can detect and identify the substance's characteristic odor.
OSHA	The Occupational and Safety Health Administration is a federal or state agency under the Department of Labor that publishes and enforces safety and health regulations for most businesses and industries in the United States.
OSU HAZCOMM	OSU Environmental Health Services Department Hazard Communications Section
OSU HAZMAT	OSU Environmental Health Services Department Hazardous Materials Section
Oxidizer	A substance such as chlorate, permanganate, inorganic peroxide, nitrocarbonitrate, or a nitrate that yields oxygen readily to stimulate the combustion of organic matter.
Oxygen Deficiency	An atmosphere having less than the normal percentage of oxygen found in normal air. Normal air contains approximately 21% oxygen at sea level.
PEL	See Permissible Exposure Limit.
Permissible Exposure Limit	An exposure limit that is published and enforced by OSHA as a legal standard. PEL may be either a time-weighted-average (TWA) exposure limit (8 hour), a 15-minute short term exposure limit (STEL), or a ceiling (C). The PELs are found in Tables Z-1, Z-2, or Z-3 of 29 CFR 1910.100. This level of exposure is deemed to be the maximum safe concentration and is generally the same value as the threshold limit value (TLV).
Personal Protective Equipment	Any devices or clothing worn by the worker to protect against hazards in the environment. Examples are respirators, gloves, and chemical splash goggles.
Physical Hazard	A substance which is a compressed gas, explosive, flammable, organic peroxide, oxidizer, pyrophoric, unstable or water reactive.

POLYMERIZATION   A CHEMICAL REACTION IN WHICH TWO OR MORE SMALL MOLECULES COMBINE TO FORM LARGER MOLECULES THAT

CONTAIN REPEATING STRUCTURAL UNITS OF THE ORIGINAL MOLECULES. A HAZARDOUS POLYMERIZATION IS THE ABOVE REACTION WITH AN UNCONTROLLED RELEASE OF ENERGY.

Reactivity   A substance's susceptibility to undergoing a chemical reaction or change that may result in dangerous side effects, such as explosion, burning, and corrosive or toxic emissions. The conditions that cause the reaction, such as heat, other chemicals, and dropping, will usually be specified as "Conditions to Avoid" when a chemical's reactivity is discussed on a MSDS.

Respirator   A device which is designed to protect the wearer from inhaling harmful contaminants.

Respiratory Hazard   A particular concentration of an airborne contaminant that, when it enters the body by way of the respiratory system or by being breathed into the lungs, results in some bodily function impairment.

Sensitizer   A substance that may cause no reaction in a person during initial exposures, but afterwards, further exposures will cause an allergic response to the substance.

Sharps   Hypodermic needles, syringes, (with or without the attached needle), pasteur pipettes, scalpel blades, suture needles, blood vials, needles with attached tubing, and culture dishes (regardless of presence of infectious agents). Also included are other types of broken or unbroken glassware that were in contact with infectious agents, such as used slides and cover slips.

Short Term Exposure Limit   Represented as STEL or TLV-STEL, this is the maximum concentration to which workers can be exposed for a short period of time (15 minutes) for only four times throughout the day with at least one hour between exposures.

"SKIN"   This designation sometimes appears alongside a TLV or PEL. It refers to the possibility of absorption of the particular chemical through the skin and eyes. Thus, protection of large surface areas of skin should be considered to prevent skin absorption so that the TLV is not invalidated.

STEL   See Short Term Exposure Limit

Synonym   Another name by which the same chemical may be known.

Systemic   Spread throughout the body; affecting many or all body systems or organs; not localized in one spot or area.

Teratogen   An agent or substance that may cause physical defects in the developing embryo or fetus when a pregnant female is exposed to that substance.

Threshold Limit Value   Airborne concentrations of substances devised by the ACGIH that represents conditions under which it is believed that nearly all workers may be exposed day after day with no adverse effect. TLVs are advisory exposure guidelines, not legal standards, that are based on evidence from industrial experience, animal studies, or human studies when they exist. There are three different types of TLV's: Time Weighted Average (TLV-TWA), Short Term Exposure Limit (TLV-STEL) and Ceiling (TLV-C). (See also PEL.)

Time Weighted Average   (TLV-TWA, Threshold Limit Value-Time Weighted Average) The time weighted average airborne chemical concentration for a normal eight hour work day and a 40 hour work week to which nearly all workers may be repeatedly exposed, day after day, without adverse effect.

TLV   See Threshold Limit Value.

Toxic   Substances such as carcinogens, irritants, or poisonous gases, liquids, and solids which are irritating to or affect the health of humans.

Toxicity   The potential of a substance to exert a harmful effect on humans or animals and a description of the effect and the conditions or concentrations under which the effect takes place.

Trade Name   The commercial name or trademark by which a chemical is known. One chemical may have a variety of trade names depending on the manufacturers or distributors involved.

TWA   See Time Weighted Average.

UEL   See Upper Explosive Limit.

UFL   See Upper Explosive Limit.

Unstable Liquid   A liquid that, in its pure state or as commercially produced, will react vigorously in some hazardous way under shock conditions (i.e., dropping), certain temperatures, or pressures.

Upper Explosive Limit   Also known as Upper Flammable Limit. Is the highest concentration (expressed in percent of vapor or gas in the air by volume) of a substance that will burn or explode when an ignition source is present. Theoretically above this limit the mixture is said to be too "rich" to support combustion. The difference between the LEL and the UEL constitutes the flammable range or explosive range of a substance. (See also LEL.)

Vapor   The gaseous form of substances which are normally in the liquid or solid state (at normal room temperature and pressure).

Water Reactive   Substances that react violently when in contact with water. They can be either be flammable solids or corrosives.